Cambridge Elements

Elements in the Philosophy of Mathematics
edited by
Penelope Rush
University of Tasmania
Stewart Shapiro
The Ohio State University

DEFINITIONS AND MATHEMATICAL KNOWLEDGE

Andrea Sereni
Scuola Universitaria Superiore IUSS Pavia

CAMBRIDGE
UNIVERSITY PRESS

Shaftesbury Road, Cambridge CB2 8EA, United Kingdom

One Liberty Plaza, 20th Floor, New York, NY 10006, USA

477 Williamstown Road, Port Melbourne, VIC 3207, Australia

314–321, 3rd Floor, Plot 3, Splendor Forum, Jasola District Centre,
New Delhi – 110025, India

103 Penang Road, #05–06/07, Visioncrest Commercial, Singapore 238467

Cambridge University Press is part of Cambridge University Press & Assessment,
a department of the University of Cambridge.

We share the University's mission to contribute to society through the pursuit of
education, learning and research at the highest international levels of excellence.

www.cambridge.org
Information on this title: www.cambridge.org/9781009517676

DOI: 10.1017/9781009091084

First published 2024

A catalogue record for this publication is available from the British Library.

ISBN 978-1-009-51767-6 Hardback
ISBN 978-1-009-09599-0 Paperback
ISSN 2399-2883 (online)
ISSN 2514-3808 (print)

Cambridge University Press & Assessment has no responsibility for the persistence
or accuracy of URLs for external or third-party internet websites referred to in this
publication and does not guarantee that any content on such websites is, or will
remain, accurate or appropriate.

Definitions and Mathematical Knowledge

Elements in the Philosophy of Mathematics

DOI: 10.1017/9781009091084
First published online: December 2024

Andrea Sereni
Scuola Universitaria Superiore IUSS Pavia

Author for correspondence: Andrea Sereni, andrea.sereni@iusspavia.it

Abstract: This Element discusses the philosophical roles of definitions in the attainment of mathematical knowledge. It first focuses on the role of definitions in foundational programs, and then examines their major varieties, both as regards their origins, their potential epistemic roles, and their formal constraints. It examines explicit definitions, implicit definitions, and implicit definitions of primitive terms, these latter being further divided into axiomatic and abstractive. After discussing elucidations and explications, various ways in which definitions can yield mathematical knowledge are surveyed.

Keywords: mathematical definitions, axioms, abstraction principles, foundations of mathematics, mathematical knowledge

ISBNs: 9781009517676 (HB), 9781009095990 (PB), 9781009091084 (OC)
ISSNs: 2399-2883 (online), 2514-3808 (print)

Contents

1 Introduction

Since antiquity, definitions have played a crucial role in the organization of our rational inquiries. Many of Plato's dialogues, to mention a familiar paradigm, can be seen as excruciating quests for definitions. The more definite the language and subject matter of our inquiries are, the more rigorous the treatment of definitions will be. Mathematics is thus a privileged arena. Given the exactness that mathematical contexts require, the study of definitions reached a mature stage once formal tools became sophisticated enough. Hence, most of the following discussion is indebted to the advancements that have occurred since the mid nineteenth century in mathematics, logic, and philosophy. The reliance on formal tools should not be overestimated though, for many theoretical features of definitions bear relevance to mathematical practice even before and beyond formalization.

As Euclid's *Elements* and Aristotle's *Posterior Analytics* emphasized, definitions are essential to any systematic inquiry. They are central to any project of conceptual analysis, especially to foundational projects. If these are seen as aiming at the attainment and organization of knowledge, definitions take on a genuine epistemological function. This Element focuses on the ways in which definitions may constitute, provide, or otherwise lead to mathematical knowledge.

What follows is thought of as an initial guide to a vast debate, which has nevertheless rarely been given a self-standing treatment (notable exceptions are Robinson 1950; Suppes 1957, chapter 8; Dubislav 1981; Belnap 1993; Antonelli 1998; Gupta and Mackereth 2023). We first rehearse the role of definitions in foundational projects (Section 2). We then discuss three major kinds of definitions: explicit definitions (Section 3), implicit definitions (Section 4), and implicit definitions of primitive terms (Section 5), the latter being divided into axiomatic (Section 5.1) and abstractive (Section 5.2). After pausing on the notions of elucidation and explication (Section 6), we eventually survey (Section 7) a variety of epistemological issues concerning definitions. We'll look for a balance among historical context, formal tools, and philosophical investigations, assuming some background but with the inexperienced reader in mind. Most mathematical examples will be confined, for analogous reasons, to geometry, arithmetic, and analysis. A consistent part of our discussion will concern authors who pioneered the foundations of mathematics (prominently Dedekind, Frege, Hilbert, Peano, Russell, and Carnap), and proponents of major contemporary views. Logicist and structuralist views, and a comparison between the two, will be given special attention, due to their role in foundational debates and their reliance on two major sorts of definitions for mathematical primitives.

2 Definitions and Foundations

2.1 Foundations

A vivid outline of a foundational project for mathematics is offered by Quine (1969), 69–70:

> Studies in the foundations of mathematics divide symmetrically into two sorts, conceptual and doctrinal. The conceptual studies are concerned with meaning, the doctrinal with truth. The conceptual studies are concerned with clarifying concepts by defining them, some in terms of others. The doctrinal studies are concerned with establishing laws by proving them, some on the basis of others. Ideally the more obscure concepts would be defined in terms of the clearer ones so as to maximize clarity, and the less obvious laws would be proved from the more obvious ones so as to maximize certainty. Ideally the definitions would generate all the concepts from clear and distinct ideas, and the proofs would generate all the theorems from self-evident truths.

Deductive *proofs* transfer (truth, justification, and) certainty from basic truths to theorems, while *definitions* transfer (meaning and) clarity from basic to derivative concepts. This marks a distinctively epistemological enterprise, echoing broader projects in the foundation of knowledge – Descartes' on the rationalist side (see e.g. the *Discourse on Method*, Descartes 1637), or Hume's on the empiricist side (see e.g. Hume 1739–40). More generally, it shares elements with a traditional model for the systematization of a deductive science (de Jong and Betti 2010), starting from Aristotle's *Posterior Analytics*. But what exactly is an epistemic foundation meant to provide?

On a strong reading, foundations deliver knowledge in the subject matter of a target domain D that was previously precluded. A notable drawback is that before foundations no one could genuinely be credited with knowledge of D. If we follow a (debatable, but still traditional) tripartite definition of (propositional) knowledge as justified true belief, then since foundations secure true beliefs, of which some, at least, were antecedently possessed, such beliefs can fail to count as knowledge only because they were not justified either. Only a foundation establishes that certain p's in D are true, and provides reasons to justifiedly believe in any (true) p in D. Such strong reading may have been underlying projects like Descartes', where hyperbolic doubt challenges both truth and justification for all beliefs in order to attain those we cannot possibly doubt; in our present case, the implication that mathematical beliefs could not justifiedly be held true until the nineteenth century is obviously unpalatable.

On a more indulgent reading, foundations have an *architectural* purpose: their aim is not (just) to question whether target statements are true, or whether we have any good reasons to believe that they are, but to establish conclusively *why* they are true, to determine the ultimate grounds their truth depends on. This approach often looks back at Euclid's *Elements* as a paradigm. Descartes himself (*Discourse*, II, 19) was moved by similar thoughts: "Those long chains of utterly simple and easy reasonings that geometers commonly use to arrive at their most difficult demonstrations had given me occasion to imagine that all the things that can fall within human knowledge follow from one another in the same way […]."

At the beginning of the modern debate on the foundation of mathematics, Frege (1884) (along with Bolzano, Dedekind, and others, with due differences) endorsed similar views:

> The aim of proof is, in fact, not merely to place the truth of a proposition beyond all doubt, but also to afford us insight into the dependence of truths upon one another. After we have convinced ourselves that a boulder is immovable, by trying unsuccessfully to move it, there remains the further question, what is it that supports it so securely? (§2)

Rather than leading from lack to possession of knowledge, then, foundations may be concerned with different kinds of justification. They would replace a weaker, defeasible, possibly even *a posteriori* and inductive justification based on successful applications, with the incontrovertible justification provided by an explanation of the deductive relations connecting basic principles and theorems. This architectural approach can then be accompanied, or even motivated, by purely mathematical concerns as to how a mathematical theory is to be best systematized.

On the epistemological side, foundations require establishing how basic principles are themselves justified, or otherwise warranted, in a noninferential way. On pain of regress, as Aristotle made clear, inferential justifications must come to an end. Definitions must come to an end too. Basic terms cannot be defined from more elementary ones, and still their meaning must be available somehow. Foundations seem bound to feature both *unprovable principles* and *indefinable primitives*. Both issues are especially pressing in mathematics, whose truths and objects, on most conceptions, are inaccessible empirically or extratheoretically.

Within a foundational project, definitions can provide, or sustain, different varieties of epistemic achievements. Surely they afford an *understanding* of the meaning of the linguistic items that are being defined, or the concepts they express, required to master them competently. They can sustain *propositional knowledge* (knowledge that something is the case) with respect to both the statements which are used to lay down the definition itself, and the theorems that can be derived thanks to its introduction. They may even lead to *objectual knowledge*, namely knowledge of the objects (if any, and of any variety) of a mathematical theory. If at least some of the basic concepts F of a mathematical theory M are concepts under which objects supposedly fall, by determining when an object x is to fall under F a definition will show how to individuate objects of kind F, and the truth of M will entail countenancing their existence. In this respect, a foundation is also an ontological, and possibly metaphysical, enterprise. Its aim is to determine, according to a given theory, what kinds of objects there are and what kinds of objects they are (what their nature, or essence, is). It can also determine what kinds of objects there aren't, if they can be reduced to, or identified with, objects of other kinds. Accordingly, although our main focus will be on epistemology, matters in both semantics and ontology will also play a relevant role.

Foundationalism can come in many varieties (Shapiro 2004), some of which emphasize the theoretical and scientific significance of systematizing, unifying, and connecting mathematical theories over traditional epistemological concerns with evidence, knowledge, and justification. Moreover, even granting that foundations are possible at all, foundationalism does not exhaust the philosophy of mathematics, nor, therefore, the epistemology of definitions. Various issues concerning definitions arise when we look at phenomena like mathematical explanation or understanding, when we consider why mathematicians redefine already established notions, or when we discuss whether one definition is more natural than another; these issues are partly independent of foundational projects, may be influenced by cognitive, sociological, and pragmatic factors, and are elicited by the study of past and current mathematical practice (see e.g. Tappenden 2008; Frans, Coumans, and de Regt 2022; Coumans 2024). Some of these will cursorily surface in what follows, although this more nuanced investigation of definitions will be kept in the background while we attempt to outline and systematize a debate which is closer to traditional foundational concerns.

But what are definitions, and what guides their formulation? Definitions are statements expressing, or establishing, a relation between (the meaning of) some linguistic items and (the meaning of) other linguistic items. They establish that the meaning of a hitherto undefined symbol or expression of a given grammatical type, the *definiendum*, is to be determined on the basis of the meaning of a combination of one or more previously known symbols or expressions, the *definiens*. Relevant grammatical types in natural language are singular terms, including proper names, and predicates, including relational expressions. Within a formal system, relevant syntactic types are constants, function symbols (including term-forming operators), and predicate and relational symbols. The way in which such a determination of meaning is effected (e.g. whether there is some sort of semantic equivalence between *definiens* and *definiendum*, whether both have to be of the same syntactic or grammatical type, and so on) varies, as we will see, according to the specific kind of definition considered.

Formally, the effect of the definition is to assign the defined expression a suitable syntactic and inferential behavior. Semantically, to a very first approximation, it is to provide the *definiendum* with a set of one or more (individually necessary and jointly sufficient) defining conditions determining its semantic interpretation (usually, the individual they refer to for constants and singular terms, the function they determine for functional symbols, or the set of individuals or *n*-tuples of such individuals they denote for predicate and relation symbols). On some conceptions, definitions also target extralinguistic items by individuating the essence, nature, or constitutive features of some object or property.

Some definitions can be seen as arbitrary stipulations that a certain expression is to be given a certain meaning. In the most interesting cases, however, definitions are guided by pretheoretical patterns of use of informal notions that they aim at capturing (wholly or partially) and systematizing. This raises the question of how to establish whether a definition is successful (and what it is, in general, for a definition to be successful). Usually, this involves a process of *reflective equilibrium* (Goodman 1954/1983) between evidence and theory: we systematize patterns of use through defining conditions, then check whether the definition thus obtained is too strict (leaves out cases we would like it to cover) or too loose (it applies to cases we wouldn't want it to apply to), and then go back to adjusting the defining conditions so as to make the definition more adequate and precise. The details and significance of this procedure depend, as we will see, on different conceptions of conceptual analysis.

2.2 Names and Things

Traditionally, a distinction is drawn between *nominal* and *real* definitions, that is, definitions *of names* (*quid nominis*) versus definitions *of things* (*quid rei*). The distinction is examined in Aristotle's remarks on definitions (Deslauriers 2007), especially in *Posterior Analytics* (CW, I), and has been later related to definitions in Euclid's *Elements* (e.g. by Saccheri; see Heath's commentary to Euclid 1926, I). To a first approximation, nominal definitions target linguistic items, that is, they provide linguistic expressions with meaning (hence, the concepts they express with content); real definitions target the objects themselves and capture their essential properties.

On one reading of Aristotle, real definitions differ from nominal ones because they reveal *why* an object is (what is the cause, or *aitia*, of their existence) and give "a demonstration of the essence" (*Posterior Analytics*, II, 94a) of a thing. They provide a metaphysical explanation. On a weaker reading, real definitions are nominal definitions together with the assumption that what is being defined exists: so, while nominal definitions just answer the question *what* a thing is (*to esti*) with no commitment to its existence, other definitions also presuppose an answer to the question *whether* a thing is (*ei esti*) and claim that it is (*oti esti*). As such, they are especially suited for primitive notions, the existence of whose referents cannot be proved within the theory and must hence be presupposed. Nominal definitions simpliciter pertain to derivative notions, the existence of whose objects has to be proved from general principles common to all sciences (common axioms, *koinà axiomata*) and from the postulates and primitive notions of the relevant theory:

> Now the things peculiar to the science, the existence of which must be assumed, are the things with reference to which the science investigates the essential attributes, e.g. arithmetic with reference to units, and geometry with reference to points and lines. With these things it is assumed that they exist and that they are of such and such a nature. . . . But, with regard to their essential properties, what is assumed is only the meaning of each term employed: thus arithmetic assumes the answer to the question what is (meant by) 'odd' or 'even', 'a square' or 'a cube,' and geometry to the question what is (meant by) 'the irrational' or 'deflection' or (the so-called) 'verging' (to a point); but that there are such things is proved by means of the common principles and of what has already been demonstrated. (*Posterior Analytics* I, 10, 76b3–76b9)

For instance, geometers must assume that there are points, but they define 'triangle' nominally, leaving to demonstrations or geometrical constructions the task of proving that triangles exist. As regards primitives, or "immediates," Aristotle further writes:

> Of some things there is something else that is their explanation, of others there is not. Hence it is clear that in some cases what a thing is is immediate and a principle; and here one must suppose, or make apparent in some other way, both that they are and what they are (which the arithmetician does; for he supposes both what the unit is and that it is); but in those cases which have a middle term [in a syllogism] and for which something else is explanatory of their substance, one can, as we said, make them clear through a demonstration, but not by demonstrating what they are. (*Posterior Analytics* II, 9, 93b22–93b28)

Definitions of immediates are counted (with common axioms and postulates) among the *archai*, the first indemonstrable principles of a deductive science. What distinguishes immediates from other notions is that an explanation of why they are – their *aitia* – is not "something else": they cannot be explained by reduction to simpler items. Here nominal definitions come to an end, and the content of immediates must be made "apparent in some other way," for they provide the very subject matter of the theory itself, its *genus* (*genos*).

The distinction between nominal and real definitions has been preserved and discussed in the history of philosophy (see e.g. Locke 1690, III, vi) and has elicited varied responses. Within the analytic tradition, due to its inherent skepticism toward traditional metaphysics, real definitions have been dismissed. To wit, in discussing Aristotle on definitions, Russell (1945, 223–225), contends that the notion of essence, like that of substance, is "a hopelessly muddle-headed notion ... a metaphysical mistake, due to transference to the world-structure of the structure of sentences composed of a subject and a predicate." Other earlier reactions were friendlier. Mill (1843, I, VIII, 7), for instance, acknowledged that "definitions, though of names only, must be grounded on knowledge of the corresponding things [...] How to define a name [...] may involve considerations going deep into the nature of the things which are denoted by the name. Such, for instance, are the inquiries which form the subjects of the most important of Plato's Dialogues."

Aristotle's concern for immediates as providing the *genus* of a theory is indeed due to his reaction to Plato's account of definitions. Many Platonic dialogues are driven by questions of the "What is X?" form. These are best seen as requests for real definitions of the essence of X (love, knowledge, etc.), rather than nominal definitions of X-terms. In the *Sophist* (see also *Phaedrus*, 265d–266b), Socrates applies the so-called *method of collection and division*: to find a correct definition of X, the inquirer should first collect different examples of what we take to be X, consider them as falling under some broader kind, and then proceed, by a dichotomic process, by dividing the largest kind into two, locating our target X into one smaller kind, to be then further divided, and so on. The essence of the species (*eidos*) X is then individuated by giving its *genus* and its *differentia* (*diaphora*), which distinguishes it from other species under the same *genus*. This decompositional analysis of concepts distinguishes proper definitions from lists of cases. When asked "What is knowledge?," the young mathematician Theaetetus (*Theaetetus*, 146c–d) initially offers a list of familiar examples: geometry, cobblery, carpentry, and "the skills that belong to other craftsmen." Socrates is adamant that this is not what he is looking for. He asks for what is common to all these examples and requires a general *criterion* for establishing whether some yet unencountered item is or isn't a case of knowledge. In contemporary terminology, this requires a definition of a concept X to provide a set of *individually necessary* and *jointly sufficient* conditions for the application for X, and, if it is a concept of objects, to determine when two objects falling under the concept are distinct (conditions of identity), and whether an object falls under the concept or not (conditions of application).

Aristotle does not reject definitions of this kind, but doubts we can rely on them only. One objection is that the process of division by *genera* assumes what it wants to prove – that is, assumes the existence of the objects that we subsume progressively into broader *genera*. On the contrary, many definitions can take the form of conclusions of syllogistic reasonings: since definitions are a "demonstration of essences," their statement can occur as the conclusion of arguments (*Posterior Analytics*, II.3–10). This other form of definition is for Aristotle much more suited to a general method of a demonstrative science, while "division by genera is a small part" of such a method.

Although mathematical definitions are most of the time treated as nominal, they are still often (implicitly or explicitly) treated as real definitions and may be conceived as ways to capture the nature of mathematical objects or to determine the conditions for their existence (see Section 7.4.2).

2.3 The Euclidean Paradigm

Both Plato's and Aristotle's views on definitions were plausibly influenced by Greek geometry, and themselves have influenced the reception of its later presentation in Euclid's *Elements*. Since the latter have long been a paradigm of foundational theories, recalling even a much simplified outline will help (for more, see Euclid 1926, introduction; Mueller 1981).

Book I of the *Elements* (in Heidberg's edition) contains twenty-three definitions (ὅροι), five postulates (αἰτήματα), and five common notions (κοιναὶ ἔννοιαι). A sample of the first seven definitions (with emphasis added on the term being introduced) is:

EUCLID'S DEFINITIONS (SAMPLE)

D1 A **point** is that which has no part.

D2 A **line** is breadthless length.

D3 The extremities of a line are points.

D4 A **straight line** is a line which lies evenly with the points on itself.

D5 A **surface** is that which has length and breadth only.

D6 The extremities of a surface are lines.

D7 A **plane surface** is a surface which lies evenly with the straight lines on itself.

In Aristotle's terms, we find both immediates and nominal definitions of derivative notions. 'Point,' 'line,' and 'surface' are among the former. They identify the subject matter (*genus*) of the theory, whose existence – according to Aristotle – must be posited, and whose explanation does not depend on "something else." Other notions are derived from them. For instance, D7 defines 'plane surface' through the already available notions of surface and straight line, the latter being itself introduced in D4 in terms of line and point. Remaining expressions – 'part,' 'breadth,' 'extremities,' 'inclination,' 'meet,' 'lies evenly,' 'containing', and so on – may be seen as part of an antecedently shared language (note that extremities could be taken as primitives and used to define other notions in D3 and D6, although this would violate Aristotle's concern not to "explain the prior by the posterior"; Aristotle CW, *Topics*, I, 4, 141b.15 ff.). Following this taxonomy, statements introducing primitives (D1, D2, etc.) would not properly count as definitions, but rather illustrations or elucidations (Section 6.1) of notions whose basic grasp is antecedently and pretheoretically guaranteed (for instance by spatial intuition). So much so that they are actually never used in proofs in the *Elements*, a feature that takes them apart from other proper definitions (and raises both exegetical issues on the composition of Euclid's work – Russo 1998 – and conceptual questions as to their theoretical role as purported definitions – see Section 3.3).

The five postulates are:

EUCLID'S POSTULATES

P1 To draw a straight line from any point to any point.

P2 To produce a finite straight line continuously in a straight line.

P3 To describe a circle with any centre and distance.

P4 That all right angles are equal to one another.

P5 That, if a straight line falling on two straight lines make the interior angles on the same side less than two right angles, the two straight lines, if produced indefinitely, meet on that side on which are the angles less than the two right angles.

Postulates are not proved from anything else, and are supposed to use only primitive notions and notions already defined through them (e.g. 'straight line' is defined in D4 via D2 and D1). Differently from modern presentations of mathematical axioms, some of Euclid's postulates (e.g. P1–P3) are not properly assertions (declarative descriptions of geometrical facts supposed to hold), but rather prescriptions or instructions on how certain figures can be constructed (through ruler and compass). Also, we now call 'axioms' the principles of a theory, but ancient usage displays a subtle variety of uses for the term (Euclid 1926, Introduction, IX, §3).

Definitions and Postulates are supplemented with common notions:

EUCLID'S COMMON NOTIONS

C1 Things which are equal to the same thing are also equal to one another.

C2 If equals be added to equals, the wholes are equal.

C3 If equals be subtracted from equals, the remainders are equal.

C4 Things which coincide with one another are equal to one another.

C5 The whole is greater than the part.

These are general principles concerning quantity and magnitude, and offer the theoretical scaffolding for a derivation of geometrical theorems. Notice that proofs in the *Elements* heavily rely on actually constructed diagrams of geometrical figures. A proper separation between their propositional and visual aspects may be hard to draw and diagrams may contribute to the justification of theorems as ingredients of their derivation (Giaquinto 2007; Manders 2008).

Several other definitions of great mathematical value are found in the later books. Worth mentioning are at least those concerning ratios of magnitudes in Book V, systematizing Eudoxus' theory of proportions, which will underlie later characterizations of real numbers; and those, in Books VII–IX, illustrating the notion of unit ("that by virtue of which each of the things that exist is called one") and defining number as "a multitude composed of units" (also introducing other notions like multiple, even, odd, prime, etc.). Possibly because arithmetic, on this conception, requires no proper construction procedures (apart from the addition of one unit to other units), no proper arithmetical postulates are provided, so no axiomatic treatment of arithmetic is advanced. This was to change only in the nineteenth century.

This brief outline not only shows that Aristotle's taxonomy of definitions applies (being itself inspired by geometrical practice) to the definitions in the *Elements*. It also shows – once significant idealizations or simplifications are conceded – why the *Elements* have been the paradigm model of foundations, where certain truths lead to theorems via proof, and clear notions lead to derivative ones via definition. Whether such a paradigm is still adequate for modern conceptions of mathematical knowledge is something we will touch upon and that can variously be disputed (Paseau and Wrigley 2024). Finally, it emphasizes the irreducible role of postulates – which must be noninferentially justified – and primitives – whose meaning cannot be given in simpler terms.

3 Explicit Definitions

The previous section foreshadowed a distinction between derivative notions and indefinable primitives. Definitions of derivative notions were traditionally modeled on definitions *per genus et differentiam*. Modern treatment rather distinguishes two major varieties of definition, explicit and implicit. Here we will discuss the former.

3.1 Outline

In natural language, explicit definitions often correlate *definiens* and *definiendum* by means of the verb 'to be.' A traditional – nonmathematical – example is the definition of 'Human' as 'Rational Animal' effected by saying that to be human *is* to be animal and rational. These uses of the verb 'to be' must be distinguished not only from its uses as a copula ("2 is a prime number"), but also from uses in (formal or informal renditions of) identity statements which are not definitions ("2 is (equal to) $\sqrt{4}$"). For our purposes, those uses which are relevant in the context of definitions will be interpreted as shorthands for logical expressions like 'is identical to' ('=') or, on some occurrences, 'if and only if' (*iff*, '$\leftrightarrow$').

The preceding definition can then be seen as elliptic for the quantified statement: "For any x, x is Human *iff* x is Animal and x is Rational." If generalizable, this obviates to the need for definitions as identities, reducing all definitions to stipulations that a sentence in which the *definiendum* occurs is equivalent to a given condition free of it: $\forall x(Human(x) \leftrightarrow (Animal(x) \,\&\, Rational(x)))$. Whether this generalization is available depends on the formal language adopted. Consider the following example (Suppes 1957).

Take an arithmetical language having as primitive symbols two individual constants, '0' and '1', the operation symbols '+' and '×', and the relation symbol '<'. Assuming the meaning of these symbols as known, we could define '2' by the following stipulation: $2 = 1 + 1$. We could turn this definition by identity into a biconditional definition as follows: $2 = y \leftrightarrow y = 1 + 1$. If we drop '0' and '1' from our primitives, we could still define them via the following conditions: $0 = y \leftrightarrow \forall x(x + y = x)$, $1 = y \leftrightarrow \forall x(x \times y = x)$. However, we could not define them via identities: there would be nothing to be added to anything in order to stipulate that $0 = x + y$; and in stipulating that $1 = 0 + 1$ the *definiendum* would circularly occur in the *definiens*. Circumventing these limitations requires complicating the background language, for example by including a description operator 'the object x such that …', ('$\daleth\, x$'), and then defining '0' by identity as follows: $0 = (\daleth\, x)[\forall x(x + y = x)]$. Alternatively, one can take '0' as the only primitive constant, and add the successor function,

$Suc(x)$, as primitive, defining every numerical constant from '0' and successor: $1 = Suc(0); 2 = Suc(Suc(0))$, and so on. As this simple example makes clear, the availability of one or another kind of definition for the same symbol may depend on the logic underlying the formal rendition of a theory and on a specific choice of primitive symbols. More generally, there is a strict interconnection between what is considered as logical and adopted as a background language in which a mathematical theory is cashed out, what mathematical notions can be defined, and how they can be defined. We'll see some more familiar examples in discussing Peano Arithmetic in Section 5.1.1.

As a general characterization of explicit definitions we follow Boolos, Burgess, and Jeffrey (2007), 266 (see also Lésniewski 1931; Suppes 1957; Antonelli 1998; Gupta and Mackereth 2023; Šikić 2022; for a model-theoretic characterization, see Hodges 1993, 302). Where 'α' and '$\beta_1, \ldots, \beta_n$' are non-logical symbols of the language $\mathcal{L}$ of some theory T and α is not among the β_i, then, depending on the syntactic type:

> In the case of a $(k + 1)$-place predicate, such a definition is a sentence of the form
> $$\forall x_0, \forall x_1, \ldots, \forall x_k(\alpha(x_0, x_1, \ldots, x_k) \leftrightarrow B(x_0, x_1, \ldots, x_k))$$
> and in case of a k-place function symbol, such a definition is a sentence of the form
> $$\forall x_0, \forall x_1, \ldots, \forall x_k(x_0 = \alpha(x_1, \ldots, x_k) \leftrightarrow B(x_0, x_1, \ldots, x_k)),$$
> where in either case B is a formula whose only nonlogical symbols are among the β_i. (Constants may be regarded as 0-place function symbols [...]. In this case the right side of the biconditional would simply be $x_0 = \alpha$.) The general form of a definition may be represented as
> $$\forall x_0, \ldots \forall x_k(- \alpha, x_0, \ldots, x_k - \leftrightarrow B(x_0, \ldots, x_k)).$$

In formal systems, definitions are singled out by appropriate notational devices, tags like '(Df.)' or subscripts like '$=_{df}$'. This became standard through the Peano school (Burali-Forti 1894, 120–121). Whitehead and Russell (1910–13), I, 11, use '(Df.)' and clarify that "It is to be understood that the sign '=' and the letters 'Df' are to be regarded as together forming one symbol." Frege too adopted specific notational devices in his *Begriffschrift* – by flanking his judgment stroke ($\vdash$), which signals, in that system, that the content following it expresses a judgment, by an additional vertical stroke ($\Vdash$).

This is not merely a notational remark. If not for a given relation of priority between the symbols flanking identity or biconditional, nothing formally distinguishes definitions (e.g. '$2 = Suc(1)$') from other formulae of similar form (e.g. '$2 = \sqrt{4}$'). Crucially, nothing formally distinguishes the *definition* '$2 = 1 + 1$' from the *theorem* '$2 = 1 + 1$'. As Boolos et al. (2007), 266, claim, explicit definitions:

embod[y] the idea that a theory defines a concept in terms of others when 'a definition of that concept in terms of the others is a consequence of the theory'. […] we say that α is explicitly definable in terms of the β_i in T if a definition of α from the β_i is one of the sentences in T.

Still, something must set definitions apart from other statements. One reason for this is that the linguistic act of stipulating something is essentially different from the linguistic act of asserting that something is the case. Frege, for one, concurred:

> I should like to divide up the totality of mathematical propositions into definitions and all the remaining propositions (axioms, fundamental laws, theorems). Every definition contains a sign (an expression, a word) which had no meaning before and which is first given a meaning by the definition. Once this has been done, the definition can be turned into a self-evident proposition which can be used like an axiom. But we must not lose sight of the fact that a definition does not assert anything but lays down something. (Letter to Hilbert, 27.12.1899, Frege 1980, 36)

And he adds elsewhere that "It is absolutely essential for the rigor of mathematical investigations, not to blur the distinction between definitions and all other propositions" (Frege 1971, 24–25). The notational aids help formalizing a pragmatic ingredient that goes beyond content, namely that the relevant formulae are not being advanced as assertions, but as definitions, this being a precondition for asserting any formula in which the *definienda* occur, including the very same formulae giving their definition. If not the form, then, it is the role of definitions that singles them out. In Frege, this role also depends on, and reflects, the kind of illocutionary force with which formulae are put forward, and hence the linguistic act that is being effected. Whitehead and Russell (1910–13), I, 11, claim that "a definition […] is not true or false, being the expression of a volition, not a proposition." Defining and asserting (to mimic Austin's motto) are different ways of doing things with words.

3.2 Roles

Explicit definitions require *definiens* and *definiendum* to be semantically equivalent. As a consequence, in any statement of a theory, *definiens* and *definiendum* can be replaced without change in truth (*salva veritate*). This is limited to nonopaque contexts (context other than belief ascriptions, modal contexts, etc.) which are nevertheless negligible in the formal reconstruction of mathematical theories (although possibly not in their development, where for example establishing belief in hitherto unproved identities has a significant cognitive import).

Explicit definitions, therefore, essentially work as helpful *notational abbreviations*: they do not introduce *any* new content, nor allow the derivation of *any* new consequence, with respect to the original theory. Both Frege and Whitehead and Russell shared this view. In the Foreword to *Grundgesetze* (Frege 1893–1903, I, vi) we read:

> Definitions themselves are not creative, and in my view must not be; they merely introduce abbreviative notations (names), which could be dispensed with were it not for the insurmountable external difficulties that the resulting prolixity would cause.

Elsewhere Frege stresses that "No definition extends our knowledge. It is only a means for collecting a manifold content into a brief word or sign, thereby making it easier to handle" (Frege 1971, 24). Similarly, in the Introduction to *Principia Mathematica* (Whitehead and Russell 1910–13, I, 11) we read: "'definition' does not appear among our primitive ideas, because the definitions are not part of our subject, but are, strictly speaking, mere typographical conveniences. Practically, of course, if we introduced no definitions, our formulae would very soon become so lengthy as to be unmanageable; but theoretically, all definitions are superfluous."

The same conception is found in the Peano school (Burali-Forti 1894, 123) and has been shared subsequently. Nonetheless, there are several ways in which explicit definitions have philosophical roles to play.

3.2.1 Salience

Whitehead and Russell (1910–13), I, 11, themselves acknowledge that: "in spite of the fact that definitions are theoretically superfluous […] a definition usually implies that the definiens is worthy of careful consideration. Hence the collection of definitions embodies our choice of subjects and our judgement as to what is most important."

Different combinations of expressions can be used as *definienta* for abbreviation, and we are at liberty to pick. Often, in the course of a proof, abbreviations are given for convenience. But other abbreviations can encode salient or crucial notions. For instance, the fact that in a given set-theory we choose to use '$\{\emptyset, \{\emptyset\}, \{\emptyset, \{\emptyset\}\}\}$' as a *definiens* for '3' points to the fact that that particular expression, possibly opposed to other available choices, plays a meaningful role in our attempt to recover arithmetic set-theoretically. We can abbreviate anything, but some abbreviations are more important than others.

3.2.2 Conceptual Analysis

Salience pertains to *definienda* too. In mathematics, *definienda* will often be inherited from a varied, partly symbolic and partly informal tradition, and a possibly unsystematic practice. Explicit definitions may then be the final step of a process of conceptual analysis. Whitehead and Russell (1910–13), Introduction, 12, mention Cantor's analysis of the notion of continuum as "the statement that what he is defining is the object which has the properties commonly associated with the word 'continuum,' though what precisely constitutes these properties had not before been known." A definition aims at "making definite" (coherently with the greek term 'ὅρος', meaning 'boundary'): "it gives definiteness to an idea which had previously been more or less vague." Countless examples could be mentioned, starting from definitions of simple geometrical figures in Euclidean geometry. In general, rigorous (possibly formalized) definitions will provide informal notions with a sharp meaning by determining the extent and limits of their application. Here, *definiendum* and *analysandum* coincide: the explanatory direction of the definition within the theory, from *definiens* to *definiendum*, reverses the prior process of analysis from an insufficiently clear notion back to its basic constituents. Conceptual analysis has a long history both within and outside mathematics (Otte and Panza 1997; Beaney 2021), is a central ingredient to foundational projects, and can be given different readings, some of which will be discussed in what follows.

3.2.3 Ontological Reduction: Identification

Explicit definitions may be a tool for ontological reduction. Suppose we have a translation procedure from an *F*-theory to a *G*-theory, allowing – via some suitable mapping – to translate talk of *F*'s (e.g. natural numbers) into talk of *G*'s (e.g. sets, or categories). By itself, this only tells us that *G*-talk can be used – possibly even only for practical purposes – to express things about *F*'s. Many cases of mathematical modeling in the sciences behave this way. But one can also take the translation to provide the real content of *F*-statements, hence entailing that to be an *F* *is* to be a *G*, that we need not consider *F*'s as additional entities with respect to *G*'s, for they simply are some of the *G*'s under different names.

Part of the process of arithmetization of analysis which, during the nineteenth century, led to ultimately base analysis on arithmetic, can be interpreted along these lines. The stepwise introduction of higher number systems from lower ones mirrors a process of ontological identification of higher numbers with set-theoretical constructions out of lower ones, thus combining conceptual analysis and ontological reduction. Following Hilbert, this is called the *genetic method* – although Hilbert thought only axiomatic presentations could offer proper foundations, and attributed to this method primarily an heuristic function (Hilbert 1900; Landry 2013). A sketch of this process is given here.

DEFINING NUMERICAL SYSTEMS GENETICALLY

Let a two-place relation $R(x,y)$ be an equivalence relation if it is:

- reflexive: $R(x,x)$
- symmetric: $R(x,y) \rightarrow R(y,x)$
- transitive: $R(x,y) \wedge R(y,z) \rightarrow R(x,z)$

Given a domain D, an equivalence relation R partitions D in equivalence classes containing all and only the elements of D such that $R(x,y)$. Every object is assigned to one and only one equivalence class. An object a which does not stand in R with any other object is assigned, given reflexivity, to a class containing only itself.

Consider numerical systems $\mathbb{C}$ (complex numbers), $\mathbb{R}$ (real numbers, including irrationals), $\mathbb{Q}$ (rational numbers), $\mathbb{Z}$ (integers), and $\mathbb{N}$ (natural numbers).

Let us define the objects of $\mathbb{Q}$ in terms those of $\mathbb{Z}$. Take the equivalence $R_1 = \langle a, b \rangle \sim \langle c, d \rangle$ between two ordered pairs of integers such that $a \times d = b \times c$, b and d being different from 0. For example, $\langle 1, 2 \rangle$ and $\langle 2, 4 \rangle$ satisfy R_1, since $1 \times 4 = 2 \times 2$, and the same holds for each of them and $\langle 4, 8 \rangle$. The equivalence class containing all such pairs, $\{\langle 1, 2 \rangle, \langle 2, 4 \rangle, \langle 4, 8 \rangle, \ldots\}$, identifies (*is*) a given rational number. Intuitively, any pair is a representative of the rational whose numerator is the first element of the pair and whose denominator is the second element of the pair: $\frac{1}{2} = \frac{2}{4} = \frac{4}{8} = \ldots$. Negative rationals are obtained by taking negative integers in the relevant pairs: $\{\langle 1, 2 \rangle, \langle -2, -4 \rangle, \ldots\}$ is $-\frac{1}{2} = -\frac{2}{4}, \ldots$.

Integers in $\mathbb{Z}$ are obtained as equivalence classes of ordered pairs of naturals in $\mathbb{N}$ satisfying $R_2 = \langle a, b \rangle \sim \langle c, d \rangle$ such that $a + d = b + c$. Intuitively, each pair represents the integer obtained by subtracting the second element of the pair from the first. For instance, the class $\{\langle 2, 1 \rangle, \langle 3, 2 \rangle, \ldots\}$ identifies (*is*) the positive integer 1, while $\{\langle 1, 2 \rangle, \langle 2, 3 \rangle, \ldots\}$ is the negative integer -1.

These definitions reduce the rationals to (classes of pairs of) integers, and the integers to (classes of pairs of) naturals. Complex numbers in $\mathbb{C}$ are easily reduced to the real numbers in $\mathbb{R}$, since they can be expressed as the sum $a + ib$, the first addendum (the real part) being a real number a, and the second (the imaginary part) being the product of a real number b by the imaginary unit i (which is such that $i^2 = -1$).

The reduction of $\mathbb{R}$ to $\mathbb{Q}$ was provided in two alternative ways by Cantor (1872) and Dedekind (1872). Let us briefly sketch Dedekind's. Take the set of all rational numbers. For expository purposes, this can be pictured as a rational line (containing points constructible as ratios from a given unit length). Divide the rationals in two classes, A_1 and A_2, such that all numbers a_1 in A_1 are less than numbers a_2 in A_2. Designate such division, called a *cut* (*Schnitt*), by '(A_1, A_2)'. If the cut is produced by a rational a, this can be either the greatest number in A_1, or the smallest number in A_2. However, there are infinitely many a_i which are neither in A_1 nor in A_2. For instance, if A_1 contains all numbers whose square is less than 2, and A_2 all numbers whose square is greater than 2, (A_1, A_2) will not be a rational number and hence not an element of either A_1 or A_2 – it corresponds to $\sqrt{2}$. Cuts (A_1, A_2) which belong neither to A_1 nor to A_2 are irrational numbers. Each (term for an) irrational number can thus be explicitly defined as (the term for) the set containing the pair of sets of (infinitely many) rational numbers which are, respectively, less than and greater than the cut.

3.2.4 Ontological Reduction: Elimination

Explicit definitions can be used to show that reference to a kind of objects can be dispensed with. This can be motivated by a principle of ontological parsimony, or by epistemological concerns with the disputed objects.

One leading conception of the subject matter of mathematics is *mathematical platonism*. According to platonists, mathematical theories (their axioms and theorems) describe facts holding of domains of *sui generis* mathematical objects. These are conceived as *abstract*, that is, noncausal, nonspatiotemporally located, and possibly necessarily existent objects, whose existence is independent of the mental and linguistic activities of human subjects (Hale 1988; Dummett 1991, chapter 18; Falguera, Martínez-Vidal, and Rosen 2022). The opposing view, rejecting the existence of (some or all) abstract mathematical objects, is called (respectively local or global) *nominalism* (Burgess and Rosen 1997). Platonism immediately raises the worry of how we can have knowledge of such objects, especially if one demands such knowledge to fall within the limits of the broadly empirical ways in which we access objects in the natural world. This epistemological *problem of access* traces back to Plato's times, but was then emphasized by Benacerraf (1973) (Linnebo 2006; Liggins 2010; Panza and Sereni 2013; Linnebo 2023; Nutting 2024). It heavily hinges on the theory of knowledge being assumed in the background. Benacerraf initially endorsed a causal theory of knowledge (Goldman 1967) requiring a causal connection between a (true) belief that p and the fact that p, something which is obviously precluded for facts involving abstract objects. In an attempt to avoid reliance on any epistemological assumption, Field (1989) reformulated the challenge as the request for an explanation of mathematicians' reliability in their belief-forming processes, apparently making the challenge even stronger (but see Sereni 2016). In what follows, we'll encounter several ways in which definitions can contribute an answer to Benacerraf's challenge, by providing mathematical knowledge or facilitating assess to mathematical objects.

One way (among others) to argue for platonism goes through semantic analysis of mathematical language. To take Bencerraf's example, consider the sentences (A) 'There are at least three large cities older than New York', and (B) 'There are at least three perfect numbers greater than 17'. Provided we take the surface-grammar of sentences at face value and as conducive of their logical form, (A) and (B) share the same logical form, that is, that there are at least mutually distinct objects x, y, and z which stand in a relation R with an object a which is such that $F(a)$ and $G(a)$. Under a standard (referential, Tarskian) semantics, singular terms and constants denote first-order objects. For (A) and (B) to be true, the objects being the referent of 'a' and the values of x, y, z must exist.

The referential import of numerical expressions may not be evident at first. In the applied arithmetical statement:

There are two moons of Mars. (M^a)

'two' occurs in *adjectival* position and, as any predicate, it should denote a property (being two). However, one can follow Frege in believing that the grammatical form of M^a is misleading, its proper logical form being that of an identity statement where numerical expressions occur in *substantival* position:

The number of the concept $\ulcorner$Moons of Mars$\urcorner$ = 2. (M^s)

Since M^s, if true, entails (in a non-free logic and in negative free logic) by existential generalization that the referents of the terms flanking the identity sign exist, then – assuming numerical objects are abstract – the truth of M^s entails a form of platonism.

To avoid this conclusion, a nominalist may want to explain the meaning of M^s while doing away with reference to objects. This can be done by reverting to M^a and defining it explicitly in non-arithmetical language:

$$\exists x \exists y [M(x) \land M(y) \land x \neq y \land \forall z(M(z) \rightarrow z = x \lor z = y)].$$ (M^{nq})

M^{nq} tells us that x and y are distinct M's and anything which is an M is either x or y – expressing in non-arithmetical terms what we say arithmetically with M^s (substantivally) and M^a (adjectivally). The occurrence of 'n' in any sentence of the form "There are n F's" should be seen as a meaningless syntactic string in a complex unstructured predicate, while the whole sentence gets replaced by a corresponding application of a numerical quantifier explicitly defined in non-arithmetical language (Frege 1884, §§55–56; Field 1980/2016, 21–23), for example:

$$\exists_0 x F(x) =_{df} \neg \exists x F(x), \qquad\qquad\qquad\qquad (\exists_0^{nq})$$

$$\exists_1 x F(x) =_{df} \exists x F(x) \wedge \forall y (F(y) \rightarrow y = x). \qquad\qquad (\exists_1^{nq})$$

These *numerical quantifiers* are explicitly (and recursively) definable. Eliminative definitions of this kind restrict or simplify the language of a theory through the *elimination* of symbols, and logical analysis is used to dispense with the apparent commitments to disputed objects.

3.2.5 Reduction, Elimination, and Multiple Realizability

A challenge to using explicit definitions for ontological reduction is raised by Benacerraf (1965). Suppose we want to reduce arithmetic to set-theory, explicitly defining numerical terms via set-theoretical ones. To claim that natural numbers are sets, we must be able to tell which sets they are: given a numerical term 'n' and terms referring to distinct sets s_1 and s_2, then either $n = s_1$ or $n = s_2$, but not both. To reconstruct arithmetic, a subject will need to identify a subset of the set-theoretical universe of the same cardinality as $\mathbb{N}$ ($\aleph_0$), a set as initial element (i.e. to play the role of 0), and an injective function on sets which behaves as the successor function on naturals. This delivers a set of sets isomorphic to the natural numbers, whose sequence is called *progression*, or ω-sequence (two theories S and S^* are isomorphic if there is a bijection mapping their domains and vocabularies which is structure-preserving, that is, such that all theorems of S about S-objects are mapped, or translated, into theorems of S^* about S^*-objects, and vice versa). The subject will then need to explain how numbers can be used not only to count *intransitively*, that is, to list numerals in the correct order, but also *transitively*, that is, to state how many objects are in a given collection or set. Two procedures are available. One subject (Ernie, in Beancerraf's story) follows von Neumann, starting with $\emptyset$ and applying a function which provides sets which are the union set of their predecessors: $\emptyset, \{\emptyset\}, \{\emptyset, \{\emptyset\}\}, \{\emptyset, \{\emptyset\}, \{\emptyset, \{\emptyset\}\}\} \ldots$ Another subject (Johnny) follows Zermelo, using the singleton function instead, delivering sets which are singletons of their predecessor: $\emptyset, \{\emptyset\}, \{\{\emptyset\}\}, \{\{\{\emptyset\}\}\}, \ldots$ These sequences are provably different, since for any two nonconsecutive sets a and b, '$a \in b$' will be true for Ernie (von Neumann's sets contain all their predecessors), and false for Johnny (the unique member of singleton sets is their immediate predecessor). However, since both set-theoretical reconstructions are isomorphic to the natural numbers, arithmetic can be translated into either. But which is which? Is $2 = \{\emptyset, \{\emptyset\}\}$ or is $2 = \{\{\emptyset\}\}$? Since arithmetic itself isn't able to settle the question, we cannot say which of the two sets 2 is, hence we cannot say that 2 is a set. Numbers could not be sets. The argument works for any system of objects isomorphic to the natural numbers, hence the conclusion generalizes: numbers could not be objects. Benacerraf's conclusion (1965, 70) that arithmetic is not a study of objects, but rather the study of "the abstract structure that all progressions have in common merely in virtue of being progressions," echoes Dedekind's work (Section 5.1.1) and lies at the heart of mathematical structuralism.

One can demote this challenge (as Wright 1983, §xv, did) by seeing it as an exemplification in mathematics of the argument for the indeterminacy of translation raised by Quine (1960), chapter 2, according to which common patterns of use of a term always license different and possibly incompatible semantic interpretations: Benacerraf's challenge would not raise any specific problem with mathematics then, but would simply remind us that mathematics is, on this score, on a par with the rest of language. This notwithstanding, reduction through explicit definitions seems to require univocal identification of individual numbers, but whenever the same structure can be multiply realized, such univocal identification seems precluded, since equally effective alternatives will always be available. Elimination is obtained only in a roundabout way: the availability of incompatible but equally acceptable definitions prevents the disputed terms from standing for objects at all.

Obviously, Benacerraf's challenge relies on the assumption that the only evidence to decide between alternative constructions can be given by arithmetical considerations. But – as Benacerraf (1996) himself acknowledges – broader philosophical or theoretical considerations could be offered. These can then lead to identifying one particular construction as preferred over the others, either in set-theoretical (Section 5.1.2) or category-theoretical (McLarty 1993) terms. An alternative option is to trade the identification of individual numbers for the positing of a unique mathematical structure, this being itself an object that can be differently instantiated. As we will see, this is the strategy adopted in several structuralist views.

3.3 Criteria

When we investigate requirements for definitions, we may in principle distinguish between formal criteria – which must be met in a formal or anyway systematic theory in order for something to be a definition – and broader desiderata – which make something a good definition, or a better definition as compared to others (in a similar vein, for instance, Tarski 1933/1956, 152, distinguishes between formal correctness and material adequacy in loooking for a definition of truth). The distinction, however, is not straightforward. On the one side, certain formal criteria for a specific sort of definition may be renounced in some views; on the other, one can take it that inadequate definitions are just not definitions at all, and view also broader desiderata as necessarily required. In what follows we'll collect potential requirements for specific kinds of definitions under the general label of 'criteria' (which include both matters of formalization and matter of content and theory construction).

Let us first outline some openly formal criteria that may guide the stipulation of explicit definitions (and other kinds of definitions in some cases).

NONCIRCULARITY. A definition must not explicitly use the symbol to be defined, nor surreptitiously presuppose it, in the *definiens*. Circularity may be harder to spot when the *definiens* relies on previous definitions (in a trivial and unproblematic sense, in explicit definitions *definienda* always presupposes the meaning of *definientia* by being stipulated as having the same meaning).

SYNONYMY. *Definiens* and *definiendum* in an explicit definition must be synonymous, that is, they must have the same meaning, or at least be in some sense semantically equivalent. The tenability of synonymy (together with analyticity) has been famously challenged by Quine (1951) (see Section 5.2.2). Despite this, synonymy or some cognate kind of semantic equivalence is traditionally considered a crucial requirement for definitions, explicit ones in particular.

Spelling out semantical equivalence heavily depends on one's theory of meaning. Take for instance a framework like Frege's (after Frege 1892), where linguistic meaning has two ingredients, sense (*Sinn*) and reference (*Bedeutung*): should definitions preserve both? Among principles that should "govern the use of definitions" (Frege 1893–1903), I, §33), Frege enlists the following:

> 1. Every name correctly formed from the defined names must have a reference. Thus, for each case it must be possible to supply a name, composed of our eight primitive names, that is co-referential with it, and the latter must be unambiguously determined by the definitions [...]

Whether preservation of sense is also required, however, is controversial. In his later *Logic in Mathematics* (Frege 1979b, 208) Frege suggests that it is, as far as explicit definitions within a formal system are concerned (a different issue is whether a given analysis of an informal notion is required to preserve both; see Frege 1984, 200):

> when a simple sign is thus introduced to replace a group of signs, [...] the simple sign thereby acquires a sense which is the same as that of the group

> of signs. [...] If the definiens occurs in a sentence [...] it is true that we get
> a different sentence [...] but we do not get a different thought.

For definitions other than explicit ones (e.g. definitions by abstraction, see
Section 5.2), a difference in sense between *definiens* and *definiendum* may be
essential. Reconciling these different attitudes of Frege isn't easy (Shieh 2008).
Similar remarks apply to other semantical theories which distinguish between
extension and intension.

ELIMINABILITY. Any occurrence of the *definiendum* must always be syntactically replaceable by the *definiens* (and vice versa):

> A formula S introducing a new symbol of a theory satisfies the criterion of
> eliminability if and only if: whenever S_1 is a formula in which the new sym-
> bol occurs, then there is a formula S_2, in which the new symbol does not
> occur such that $S \rightarrow (S_1 \leftrightarrow S_2)$ is derivable from the axioms and preceding
> definitions of the theory. (Suppes 1957, 154)

Thus formulated, ELIMINABILITY is a syntactic criterion (see Gupta and
Mackereth 2023 for a semantic version). Together with SYNONYMY, it entails:

SUBSTITUTABILITY SALVA VERITATE. Any occurrence of a *definiendum* s_n in a
formula S^+ of the extended language $\mathcal{L}^+$ of a theory M^+ (obtained from M
by adding the definition) can always be replaced by its *definiens* symbols
$s_0, \ldots, s_k$ so that the formula S of $\mathcal{L}$ so obtained maintains the same truth-
value of S^+, and vice versa: S^+ is a consequence of M^+ *iff* S is. The criterion
is originally due to Leibniz. Whether substitutability *salva veritate* also entails
substitutability *salva significatione*, that is, preservation of meaning, may also
depend on whether sense (or intension) is preserved in addition to reference by
Synonymity.

SYNONYMY, ELIMINABILITY, and SUBSTITUTABILITY together entail:

NONCREATIVITY (CONSERVATIVENESS). Explicit definitions must be theoretically
dispensable: they must not introduce, or create, in M^+ consequences expressible in the language $\mathcal{L}$ of M which were not already derivable from (syntactic
version), or a consequence of (semantic version), M itself: "A formula S introducing a new symbol of a theory satisfies the criterion of non-creativity if and
only if: there is no formula T in which the new symbol does not occur such
that $S \rightarrow T$ is derivable from the axioms and preceding definitions of the theory but T is not so derivable" (Suppes 1957, 154). A semantic counterpart
of this can be obtained by replacing derivability with logical entailment (see
Gupta and Mackereth 2023). This criterion is often referred to as Conservativeness, and a theory M^+ is a *conservative extension* of M only if all definitions
adding new symbols in M^+ are noncreative.

CONSISTENCY. A definition must not make a consistent theory inconsistent. To be sure, one is free to define inconsistent notions if the abbreviations introduced follow the fate of their *definiens* (define 'Top' as 'The greatest natural number', and '$\neg \exists x \{x = Top\}$' is still a theorem of arithmetic). Now from the criteria given earlier it follows immediately that no (syntactic or semantic) inconsistency can be introduced by explicit definition if the original theory is consistent. If some of those criteria are dropped, as for some implicit definitions, consistency becomes a substantial requirement. Notice, to stress again the interdependence between mathematical theorizing and logical languages, that since CONSISTENCY concerns what follows from what, it clearly depends on how a theory's background logic interprets derivability and logical consequence.

3.3.1 Fregean Criteria

As anticipated, broader, not strictly formal, desiderata may be required of definitions. Given their historical and persisting relevance, it is worth considering some of Frege's suggestions.

FRUITFULNESS. Definitions must "show their worth" by being fruitful (*fruchtbar*). In a weak sense, this means that "it must be possible to use them for constructing proofs" (Frege 1884, xxi; also, Shieh 2008; Boddy 2021). Fruitfulness is a widely shared desideratum on definitions in mathematical practice (Tappenden 2008, 2012). Frege, in particular, connects this to the claim that "Those [definitions] that could just as well be omitted and leave no link missing in the chain of our proofs should be rejected as completely worthless" (Frege 1884, §70). Definitions that are not used anywhere in proofs are not properly definitions at all (Frege seems to have used Euclid's alleged definitions of primitives as an example – as outlined in Section 2.3 – taking them as purely "ornamental definitions"; see Frege and Carnap 2003, p. 141).

This understanding of Frege's conception, however, is controversial. After all Frege (1893–1903, I, §33) endorses both ELIMINABILITY and NONCREATIVITY, and these together guarantee that *definienda* can be eliminated anywhere without loss of content. One way of reconciling this tension is to appeal to passages where Frege suggests that definitions should provide analysis of pretheoretical concepts (e.g. natural number). The fruitfulness of a definition would then be located in the fact that the theory provides the correct *definiens* for such concepts: its absence would surely leave gaps in proofs, thereby showing the definition to be fruitful. Still, this is in tension with later works of Frege's, where the emphasis is not on conceptual reconstruction, but rather on theory construction. In *Logic in Mathematics* (Frege 1979b, 209–211), Frege distinguishes between "constructive definitions," that is, arbitrary explicit stipulations within a theory, and "analytical definitions," that is, definitions of symbols based on a logical (i.e. conceptual) analysis of previously established uses. However, there is no guarantee that such analysis will preserve meaning. Moreover, the analysis "does not [...] form part of the construction of the system, but must take place beforehand." Analytic definitions are thus no definitions at all, and "the sense in which [a] sign was used before the new system was constructed is no longer of any concern to us." In such cases, the worth of the analysis does not seem to rely on its capturing exactly the original informal meaning of prior notions, but rather on its being theoretically fruitful. As Carnap reports from Frege's lectures in Jena (Frege and Carnap 2003, p. 140), "Such an analysis cannot be proved right; one can only feel that one has hit the nail on the head, and it can prove itself fruitful." FRUITFULNESS is then strictly related to the role of **Conceptual Analysis** in theory construction, as well as to the role of elucidations and explications (see Section 6).

COMPLETENESS. Definitions must provide concepts with sharp boundaries of application. As a consequence, they cannot be expanded or modified by adapting them to new entities (see Frege 1893–1903, II, §§56 ff.). We shouldn't for example define '+' to express sum among naturals and then extend it to sum among integers, rationals, and so on, for we would either be defining a new operation, or claiming that the original was ill-defined. Frege calls such defective definitions "piecemeal": "This consists in providing a definition for a special case [...] and putting it to use and then, after various theorems, following it up with a second explanation for a different case" (Frege 1893–1903, §57). Frege acknowledges that the "development of the science which occurred in the conquest of ever wider domains of numbers, almost inevitably demands such a practice" (§58). But it would then be better to introduce new symbols, since what is acceptable *in practice* may not be acceptable *in logic*, that is, in a stable logical reconstruction of a mathematical notion relative to its full domain of application. At each step, piecemeal definitions are only shadows of definitions: "Logic cannot recognise such concept-like constructions as concepts" (§56).

Frege's insistence on completeness is motivated by his views that concepts (i.e. functions from objects to truth-values) should be defined over a universal domain, and by his adoption of the Principle of Bivalence, which demands that any sentence is either true or false, so that for every object it must be determined whether a given concept does or does not apply. These views led Frege to controversy with fellow mathematicians, especially Peano (see Frege 1893–1903, II, §58, fn. 1, 71), who believed that symbols like '+' or even '=' should be given "conditional definitions": definitions should only be given over suitably restricted domains, and Frege's idea that at some stage all such partial definitions can be "combined into the form of a single explanation" would be unreasonable.

COMPLETENESS. can be jeopardized for many ordinary concepts by vagueness, which entails an area of borderline cases in which a given concept neither clearly applies nor clearly doesn't. However, even if one believes that vagueness does not affect mathematical concepts, the criterion can be questioned on more general grounds. Some concepts may be *indefinitely extensible* (Dummett 1993c, 441), namely such that, "if we can form a definite conception of a totality all of whose members fall under [them], we can, by reference to that totality, characterize a larger [sic] totality all of whose members fall under [them]." A paradigmatic example is the concept ⌐not being a member of itself ¬ (Section 5.2.1), but other candidates are those of set, concept, extension, and ordinal and cardinal number. Indefinite extensibility may be at the origins of paradoxes afflicting these notions (Russell Paradox for sets and extensions, Burali-Forti Paradox for ordinals, etc.). To avoid paradoxes, Dummett suggests abandoning classical logic and its requirement for determinate domains of quantification, and adopting intuitionistic logic, thus forfeiting both COMPLETENESS and the Principle of Bivalence (Shapiro and Wright 2006; Uzquiano 2015; Linnebo 2018a; notice that similar conclusions can be reached in the context of intuitionist mathematics, largely inspired by Brouwer's works, where definitions must be constructive, i.e. must not merely support a proof that certain objects exist, but must allow for a finite construction procedure for them; Posy 2020).

Frege's rejection of piecemeal definitions may also be resisted by arguing that however strictly we delimit the application of a concept, there may always be new cases for which it is unclear whether the concept applies: concepts can have *open texture*, or *porosity* (Waismann 1945, 122–123). Most of our empirical concepts, for instance, "are not delimited in all possible directions" and so their definitions are "*always* corrigible or emendable." This phenomenon is different from vagueness, but still leaves a modicum of indeterminacy in our definitions, and it may affect mathematical concepts too, including most basic ones – for example, the concept of number, which groups an always corrigible "family" of more specific concepts (integer, rational, etc.) (Waismann 1951, 235–237). For any porous mathematical concept, COMPLETENESS will fail (Shapiro 2014; Shapiro and Roberts 2021).

Other challenges come from Lakatos' antifoundationalist views. Lakatos (1976) presents a dialogue between characters discussing the (Descartes–)Euler theorem for polyedra, namely that: V(ertices) − E(dges) + F(aces) = 2. In the discussion, several counterexamples are advanced. The effect of these is not just to refute the theorem, but to challenge the very initial definition of 'polyedron' on which it rests. Proofs and refutations, in mathematics, not only concern propositional content: they determine the meaning of mathematical terms (a conception evoking some Aristotelian remarks; see Section 2.2). On this picture, definitions, against COMPLETENESS, are always modifiable conjectures, part of a fallible mathematical practice (Brown 2008; De Toffoli 2021).

Notice that these challenges to COMPLETENESS have different nature. Indefinite extensibility allows for a univocal identification of the target concepts, but allocates in the very nature of those concepts their impossibility of delimiting a determinate and self-contained domain of objects. Waismann's open texture and Lakatos' views put a heavier burden on the analysis of mathematical concepts in general. They suggest that either by the underdetermined nature of concepts, or by the fallibilist nature of conceptual inquiry, a sharp determination of concepts is neither forthcoming nor probably desirable. On this score, they resonate with more nuanced conceptions of analysis, concerned with how informal notions can be differently explicated and how definitions can be modified with time for contextual or pragmatic reasons (see Section 6.2).

SIMPLICITY. A definition must "contain a single sign whose reference it stipulates," and "one must not [...] explain a sign or word by explaining an expression in which it occurs and whose remaining parts are known" (Frege 1893–1903, II, §66). Simplicity entails (a) that it is not possible to jointly define more than one symbol, and (b) that it is not possible to define a symbol α by an expression in which α itself occurs. The latter is distinct from a circularity worry and is meant to rule out methods for defining several undefined expressions at once by means of a matrix of statements in which the expressions occur. Frege (1893–1903) illustrates this with an algebraic analogy (II, §66):

> It sometimes occurs that a whole system of definitions is laid down, each containing several words that are to be explained, such that each of these words occurs in several of these definitions. This is comparable to a system of equations with several unknowns, where it is once again entirely left open to question whether there is a solution, and whether it is uniquely determined.

This requirement had been already discussed in the literature, and relates more generally to implicit definitions, to which we now turn.

4 Implicit Definitions

Despite having potentially interesting philosophical roles, explicit definitions appear limited in scope. For instance, Ayer (1936, 59–61) compares explicit definitions with those we "expect to find in dictionaries," and thus with the definitions *per genus et differentiam* of the Aristotelian tradition, and adds:

> the philosopher [...] is primarily concerned with the provision, not of *explicit* definitions, but of definitions *in use*. We define a symbol *in use*, not by saying that it is synonymous with some other symbol, but by showing how the sentences in which it significantly occurs can be translated into equivalent sentences, which contain neither the *definiendum* itself, nor any of its synonyms.

These are usually referred to as *implicit definitions*, although a closer look reveals a wide assortment of terminology, forms, and uses.

4.1 Outline

The characterization of implicit definitions traces back to Gergonne's *Essai sur la théorie des définitions* (Gergonne 1818–9). First, Gergonne introduces explicit definitions as identities as surveyed earlier (9, 13). He then claims that sometimes, even in the absence of such identities, an undefined word can be understood by understanding a sentence in which it occurs, together with (and only with) other known words. For instance, by learning that each of the two diagonals of a quadrilateral divide it into two triangles, someone familiar with 'quadrilateral' and 'triangle' may learn the meaning of the hitherto unknown 'diagonal.' These are "phrases which thus provide the understanding [*intelligence*] of one of the words of which they are composed by means of the known meaning of the others" (Gergonne 1818–9, 9, 23). Gergonne too compares explicit definitions to solved equations, and implicit definitions to unsolved equations. Crucially, the analogy also shows that "it is also conceivable that, just as two equations with two unknowns determine both of them, two sentences which contain two new words, combined with known words, can often determine their meaning; and the same may be said of a greater number of new words combined with known words, in an equal number of sentences" (Gergonne 1818–9, 9, 23).

Essential features of implicit definitions that can be retrieved from Gergonne's discussion are the following:

(G_1) CONTEXTUALITY. Symbol α is defined via a sentential context featuring α as an undefined symbol and known symbols $\beta_o, \ldots, \beta_n$ otherwise.

(G_2) SIMULTANEITY. More undefined symbols $\alpha_0, \ldots, \alpha_n$ can simultaneously be defined by a system of sentential contexts otherwise featuring only known symbols $\beta_0, \ldots, \beta_n$.

(G_3) NUMERICAL BALANCE. Each defining sentential context must contain "just one word" (14) α which is undefined.

(G_4) UNIVOCALITY. The meaning of the undefined symbol α must be completely and univocally determined through the definition.

By (G_1), implicit definitions differ from explicit ones because "the definiens will be a statement or condition that involves the term to be defined. [...] The purpose of such a definition is not, of course, to set up an equivalence between the term and the defining condition, but somehow to constrain the interpretation of the term by means of the condition" (Fine 2002, 16). Hence, α is defined only to the extent to which it contributes to the meaning of a sentence in which it occurs: "the symbol itself, when occurring in isolation, need not be endowed with any meaning at all" (Antonelli 1998, §3). Without violating circularity, therefore, the *definiens* must contain the *definiendum*. A common example is the arithmetical definition of '+' given the following conditions: $n + 0 = n$ and $n + Suc(m) = Suc(n + m)$.

Since their *definiens* is an entire sentential context, implicit definitions are often called *contextual definitions*. The expression 'definitions in use' is due to Whitehead and Russell. In the Introduction to Whitehead and Russell (1910–13), 66, they claim: "By an 'incomplete' symbol we mean a symbol which is not supposed to have any meaning in isolation, but is only defined in certain contexts. [...] Such symbols have what may be called a 'definition in use'." Their examples include mathematical symbols like '$\frac{d}{dx}$' and '$\int_a^b$', which require completion in order to mean anything, but the crucial case is the treatment of definite descriptions, namely the analysis of the expression 'the x such that $\phi(x)$' as an incomplete symbol. Despite its grammatical behavior of a singular term, its meaning is defined by a complex conjunctive condition to the extent that there exists one x, and only one x, such that $\phi(x)$.

Two classes of definitions – axiomatic and abstractive – whose primary purpose is to introduce primitive terms are usually considered implicit definitions. Indeed, they meet condition (G_1) and are thus contextual definitions (axiomatic definitions are actually rarely called contextual or in use, and the two may even be considered in opposition; see Carnap 1928, §15). However, given their role, they may forfeit some of the criteria that apply to other implicit definitions (Gupta and Mackereth 2023, §2.6, talk of a "liberalized conception of implicit definitions"). We will here distinguish between *implicit definitions* in a more general understanding of the term, and a class of *primitive implicit definitions*. To sketch a basic distinction, given a mathematical theory M with a language $\mathcal{L}$, implicit definitions proper will act, similarly to explicit definitions, as expansions of $\mathcal{L}$ to $\mathcal{L}^+$. Primitive implicit definitions, on the other hand, provide the vocabulary of $\mathcal{L}$ itself (see also Gupta and Mackereth 2023, §2.6; Šikić 2022, §6, aptly distinguishes between "definitions *in* models" and "definitions *of* models").

This section discusses only implicit definitions in general, whose formal characterization is best given in semantic or model-theoretic terms. Following again Boolos et al. (2007), 266, we can say that implicit definitions incorporate the idea that "a theory defines a concept in terms of others if 'any specification of the universe of discourse of the theory and the meanings of the symbols representing the other concepts (that is compatible with the truth of all the sentences in the theory) uniquely determines the meaning of the symbol representing that concept'."

Hence (where the β_i are among the nonlogical symbols of T) "α is implicitly definable from the β_i in T if any two models of T that have the same domain and agree in what they assign to the β_i also agree in what they assign to α."

4.2 Roles

Implicit definitions share many of their potential roles with explicit definitions. As regards **Salience**, they can still specify "our choice of subjects" via appropriate *definienda*. Also **Conceptual Analysis** of pretheoretical notions is one crucial role for implicit definitions, once a suitable primitive vocabulary is available. On this regard, they can outstrip explicit definitions when no direct semantical equivalence for an undefined symbol α is available. Recursive definitions are a common example (see Section 4.3). Both these roles, however, take up a particular philosophical significance when other purposes of implicit definitions are considered.

4.2.1 Ontological Reduction

Implicit definitions allow translating discourse about α in terms of more elementary expressions '$\beta_0, \beta_1, \ldots \beta_n$' without forcing us to identify α with one of the β_i. As Quine (1969), 72, emphasizes, even outside mathematics this procedure proved helpful, for instance in translating, in empiricist accounts of empirical knowledge talk about physical bodies into talk of sense expressions while avoiding Hume's controversial identifications of the former with the latter (Quine sees Bentham's works on fictions in Bentham 1962 as a "crucial step" in the development of this procedure).

It is to this definitional procedure that Ayer refers to in the earlier quotation, for the role it plays in Carnap's *Aufbau* (Carnap 1928), which attempts an ambitious foundation of our scientific language by means of a "constructional system of concepts." Such a system provides a "genealogy of concepts" starting from a limited basis of fundamental concepts (§1), aided by a suitable theory of relations. "Construction takes place through definition" (§38), and provides progressive "ascension" to concepts of objects of higher level from concepts of lower-level objects. In the simplest cases, explicit definitions will do, and "the new object […] remains within one of the already formed object spheres." In other cases it will only be possible to define an object "relative to the already constructed objects" (§39); Carnap's term for these is "quasi-objects." Construction requires that:

> it […] be possible to transform the propositions about it into propositions in which only the previous objects occur, even though there is no symbol for this object which is composed of the symbols of the already constructed objects. […] such an introduction of a new symbol is called a *definition in use* (*definitio in usu*), since it does not explain the new symbol itself – which, after all, does not have any meaning by itself – but only its use in complete sentences.

Carnap takes the definition of 'prime number' as an example: 'x is a prime number $\leftrightarrow x$ is a natural number and has only 1 and x as divisors'. Carnap concludes that 'prime':

> can be defined only in use by indicating which meaning a sentence of the form "a is a prime number" is to have, where a is a number. This meaning must be indicated by giving a propositional function which means the same as the propositional function "x is a prime number," and contains nothing but already known symbols, and which could thus serve as a translation rule for sentences of the form "n is a prime number."

As Carnap's terminology of "quasi-objects" suggests, ontological reduction is weaker when obtained through implicit definition: we trade identification of objects for translation of talks about them. However, this coheres with the *semantic tradition* which has characterized a large bulk of analytic philosophy (Coffa 1991; Dummett 1993b): philosophical inquiries, ontological ones included, must proceed through semantic analysis. In this setting, constructions by contextual definitions address every legitimate concern on what speaking of certain objects consists in, while all further metaphysical questions become pseudo-problems.

4.2.2 Rational Reconstruction

Carnap conceived of the *Aufbau* as a project of *rational reconstruction*, which he defines as "the searching out of new definitions for old concepts" (Carnap 1928, Preface, 2nd ed., v). Given a choice of primitives, a rational reconstruction should yield a way for a subject of acquiring *a priori* the fundamental concepts of a target discourse. This is provided by contextual definitions.

The only evidence for the correctness of a rational reconstruction is its ability to output the targeted discourse. But then nothing prevents in principle different reconstructions, based on different selections of primitives and definitions, from resulting in equally adequate outputs. As Quine (1969), 75, vividly states: "If Carnap had successfully carried such a construction through, how could he have told whether it was the right one? The question would have had no point. […] If there is one way there are many […]".

Rational reconstructions of mathematical theories by implicit definitions may face the same problem. Once a certain mathematical discourse is recovered, asking which particular reconstruction is the right one may have no point. This exacerbates the multiple realizability problem Benacerraf (1965) raises for explicit definitions.

4.3 Criteria

NONCIRCULARITY. Implicit definitions too should avoid surreptitiously relying on the meaning of the *definiendum*. This conditions may appear to be violated in cases in which the defining condition, or some statements of a set of defining conditions, involves the *definiendum* itself. A relevant case is recursive definitions, where a base clause states that a property P holds for an initial base element x, and a recursive clause establishes when, for any y obtained from x by successive application of some procedure, P holds of y. Take the definition of exponentiation in arithmetic:

$$m^0 = 1,$$
$$m^{n+1} = m^n \times m.$$

While (each instance of) 'm^n' occurs as part of the *definiens* in (each instance of) the second clause, this occurrence is harmless, since the definition abbreviates an infinite sequence of formulae in which the *definiendum* only occurs on the left:

$$m^0 = 1,$$
$$m^{1(=0+1)} = 1 \times m,$$
$$m^{2(=1+1)} = (1 \times m) \times m,$$

$$\ldots$$

When this is the case, such formulae are said to be in normal form (Gupta and Mackereth 2023, §2.4), and the conditions imposed on them guarantee satisfaction of ELIMINABILITY and NONCREATIVITY (CONSERVATIVENESS).

SYNONYMITY between subsentential expressions does not properly apply to noneliminative (see below) implicit definitions. However, some form of semantical equivalence (arguably stronger than material or logical equivalence) between the defining and the defined *sentential contexts* may still be demanded (as we'll see in the case of abstractive definitions).

CONSISTENCY. Avoidance of inconsistency is trickier than with explicit definitions, since with implicit definitions one must ascertain that the defining condition is entailed by the theory. Mates (1972), 198, exemplifies this with the definition of division on integers in terms of multiplication:

$$\forall x \forall y \forall z (x \div y = z \leftrightarrow x = y \times z).$$

This appears as a good definition, since for example $6 \div 2 = 3$ and $6 = 2 \times 3$. However, the condition also establishes that $0 \div 0 = 1$ (since $0 = 0 \times 1$) and $0 \div 0 = 2$ (since $0 = 0 \times 2$), and hence entails, inconsistently, that $1 = 2$.

NONCREATIVITY (CONSERVATIVENESS). Conservative implicit definitions added to consistent theories preempt the preceding problem: no inconsistent consequence of the extended theory would be a theorem in the language L of the original theory M. Although consistency of the extended theory is guaranteed by conservativeness (from a consistent theory), the inverse doesn't hold. A creative implicit definition may yield a consistent and yet nonconservative extension M^+ of M (see Mates 1972, 198, for examples, and the discussion of primitive implicit definitions in the next Section).

ELIMINATIVITY and SUBSTITUTABILITY SALVA VERITATE. In general, Eliminativity must be understood differently for implicit definitions than it is for explicit ones, since in implicit definitions we may not be able to replace a defined expression term by term with a defining one. Eliminativity in a broader sense may be preserved by ensuring that any *sentential context* in which the *definiendum* occurs can be replaced without loss with the defining sentential contex(s), hence guaranteeing also substitutability *salva veritate* of these entire (sets of) sentences.

In some favorable conditions, however, eliminativity can be guaranteed more straightforwardly (Shapiro 1991, §6.6; Šikić 2022; Gupta and Mackereth 2023), whenever Beth's Theorem (Beth 1953) holds. Beth builds on a method introduced by Padoa (1900), that establishes when an undefined symbol α is not definable via (is "irreducible to") a system of primitive undefinable symbols

$\beta_0, \ldots, \beta_n$ of a theory T as they occur in the axioms ("unproved propositions") of the latter:

PADOA'S METHOD

[…] to prove that the system of undefined symbols is irreducible with respect to the system of unproved propositions it is necessary and sufficient to find, for each undefined symbol, an interpretation of the system of undefined symbols that verifies the system of unproved propositions and that continues to do so if we suitably change the meaning of only the symbol considered. (Padoa 1900, 122)

Hence, α cannot be implicitly defined by the β_i in T if there are two models $\mathcal{M}^1$ and $\mathcal{M}^2$ agreeing on the interpretation of all the β_i in T but disagreeing on the interpretation of α. Now, it is a consequence of the characterization of explicit definitions and the UNIVOCALITY constraint (G_4) on implicit ones, that any α which is explicitly definable from the β_i in T is also implicitly definable from the β_i in T. Beth's Theorem conjoins this with the inverse direction (Boolos et al. 2007, §20.3; Shapiro 1991, §6.6.3; Hodges 1993, §6.6; see also van Heijenoort 1967, 118–119; Giovannini and Schiemer 2019):

BETH'S DEFINABILITY THEOREM

α is implicitly definable from the β_i in T *iff* α is explicitly definable from the β_i in T.

Beth's Theorem holds for first-order languages. In this case, any implicit definition can be turned into an explicit one, and Eliminativity is hence immediately guaranteed. However, Beth Theorem does not hold in second-order languages (due to their failing compactness; Shapiro 1991, 164; Boolos et al. 2007, 267), and many philosophically relevant implicit definitions are formulated in second-order languages. Moreover, some notable (e.g. axiomatic) first-order definitions are used to introduce primitive symbols, and may thus not be required anyway to meet all general criteria for implicit definitions, Eliminativity included. Let us then turn to primitive implicit definitions.

5 Primitive Implicit Definitions: Axioms and Abstractions

The conviction that mathematical primitive terms are undefinable was long held. To wit, in *De l'Esprit géométrique* (Pascal 1658) Pascal praises Euclidean geometry because it does not attempt to define its primitive notions, which are "clear and understood by all mankind" and "so naturally designate the things they mean, to those who understand the language," so that "there is nothing more feeble than the discourse of those who wish to define these primitive words." In contrast to similar attitudes, axiomatic and abstractive definitions have been developed as ways of defining mathematical primitives.

5.1 Axiomatic Definitions

5.1.1 Outline

In the Euclidean model, one way to picture the epistemic role of primitives and postulates is that they neither admit nor need, respectively, definition and proof, since the basic source of geometrical knowledge is immediately provided by the perceptually accessible physical space we inhabit, whose properties geometry studies (or, in a Kantian perspective, constitutes). This picture was shaken in the nineteenth century by the development of non-Euclidean geometries (mainly through the negation of the fifth postulate), which detached the study of geometrical space and physical space, and removed any kind of extratheoretical ground for postulates and primitives. Geometries started being studied for their formal properties as mathematical theories. The first axiomatic presentation of Euclidean geometry was Hilbert's, in the *Grundlagen der Geometrie* (Hilbert 1899). He considers "three distinct systems of things," points, straight lines, and planes, and proceeds to discuss geometrical axioms. Although he still presents "the choice of the axioms and the investigation of their relations to one another" as "tantamount to the logical analysis of our intuition of space," and although he presents his five groups of Axioms (connection, order, parallel, congruence, and continuity) as expressing "certain related fundamental facts of our intuition," he crucially claims that

> We think of these points, straight lines, and planes as having certain mutual relations, which we indicate by means of such words as "are situated," "between," "parallel," "congruent," "continuous," etc. The complete and exact description of these relations follows as a consequence of the *axioms of geometry* (§1, p. 3).

In order to define the geometrical primitives, it suffices to determine which relations hold between the elements points, lines, and planes, and everything there is to establish about these relations is provided by the axioms (in either Euclidean or non-Euclidean groupings). The Axioms of Connection "*establish* a connection between the concepts indicated above; namely, points, straight lines, and planes"; the Axioms of Congruence "*define* the idea of congruence or displacement"; and so on (emphasis added). Axioms do not simply capture, reconstruct, or recover relations among elements that are given to us prior to the theory, but rather establish them. While subsequent definitions are explicitly given in terms of primitives (e.g. "If M is an arbitrary point in the plane α, the totality of all points A, for which the segments MA are congruent to one another, is called a *circle*"; §7), primitives are *jointly defined by the axioms*.

Lacking extratheoretical evidence, the consistency and independence of axioms must be ascertained through appropriate mathematical procedures. To this aim, Hilbert develops (metamathematical) methods that will then constitute the bulk of model theory. To prove consistency (§9), "it is sufficient to construct a geometry where all of the five groups are fulfilled [i.e. satisfied]." Let points stand for ordered pairs $<x, y>$ from the field of real numbers (or for an appropriately selected enumerable subset Ω of algebraic numbers), and define lines and planes as constructions on them (that points, lines, surfaces could be represented as constructions out of real numbers was common since Descartes introduced analytic geometry). Then any contradiction in the axioms would surface in the theory of real (or Ω) numbers. Hilbert thus provides a model-theoretic consistency proof of his axioms relative to the consistency of analysis (or the consistency of the arithmetic of Ω), this being assumed as granted. Analogous procedures prove that axioms are mutually independent, and hence that "none of them can be deduced from the remaining ones by any logical process of reasoning" (§10).

In arithmetic, the axiomatic definition of arithmetical primitives was to be developed by Peano and had been anticipated by Dedekind (1888). Dedekind sets out to investigate "that part of logic which deals with the theory of numbers" (14). Both he and Frege saw themselves as extending to arithmetic the quest for rigor that dominated the mathematics of their time, they both set out to expel the intuitions of space and time from the foundations of mathematics, and both saw arithmetic as intimately related to logic and as, as Dedekind puts it, "an immediate result of the laws of thought" (Reck 2013, 2021; on different ways of characterizing mathematical rigor, see Burgess 2015; Tanswell 2024).

Dedekind begins by assuming the notion of an object [thing, *Ding*] as any object of thought. He then introduces the notion of a system [*System*] of objects (a proto-set-theoretical notion), having other objects as elements. He then defines subset (part), proper subset (proper part), and intersection [*Gemeinheit*]. He provides the first systematic treatment of transformations [*Abbildungen*], namely mappings between a system S and its image S', or $\phi(S)$, associating an element s of S to an element s' ($\phi(s)$) of S'. He first defines similar transformations (injective mappings, although he sometimes means bijections), and then automorphism, namely transformations of S into an S' which is a (proper or improper) subset of itself. He introduces the crucial notion of chain

[*Kette*] of a system S (S_0) as the intersection of all those systems which contain S and contain all and only those systems which can be obtained by S through ϕ. Apart from ambiguities between a system S and an element s, a chain coincides with the smallest inductive set obtained from s of S through a function ϕ (i.e. closed under ϕ). Two other major accomplishments are (a) the theorem of complete induction; (b) the first rigorous definition of an infinite set as one that stands in a bijection with a proper subset of itself (subverting a long tradition according to which the whole is greater than the part, as we read in Euclid's Common Notions). This by itself does not prove that infinite systems (sets) exist. In §66 Dedekind proves that if we take the system S of all things that can be objects of my thought, and the function ϕ mapping each element s of S to the thought that s is an object of my thought (which, being a thought, is itself an object of my thought), then the set $\phi(S)$ so obtained is a proper subset of S, and hence, by the definition of infinity, S is infinite. To prove that $\phi(S)$ is a proper subset of S, Dedekind shows that there is at least one thing which is an object of my thought (hence an element of S) but is not itself a thought of the form "s is an object of my thought" (hence not an element of $\phi(S)$). Quite unfortunately, he selects "my own Ego" as an example, thus jeopardizing his proof with psychologistic notions. The criticism of psychologism or mentalism is, however, possibly too harsh, and Dedekind's aim was in fact to provide a logical proof of the existence of infinite sets.

To define natural numbers, Dedekind introduces a specific kind of infinity, simple infinity (§71), and proves that every infinite system S contains as a (proper or improper) part a simply infinite system N (§72). Via the earlier proof that there are infinite systems, this also proves that there are simply infinite systems. These are defined by the following conditions:

DEDEKIND'S SIMPLY INFINITE SYSTEMS

α. $\phi(N)$ is a part of N
β. N is the chain obtained through ϕ from a base-element 1 ($N = 1_0$)
γ. 1 is not contained in $\phi(N)$
δ. ϕ is similar (injective)

The use of 'N' and '1' is notationally arbitrary: by themselves, conditions $\alpha - \delta$ only state that a simply infinite system S is ordered [*geordnet*] by an injective function ϕ from elements of S to elements of S, starting from an initial element s which is not in the image $\phi(S)$ of S. Natural numbers are subsequently characterized via these conditions (§73), through a process of apparently *psychological* abstraction (but see Reck 2013; Yap 2014):

> If in the consideration of a simply infinite system N set in order by a trans-formation ϕ we entirely neglect the special character of the elements; simply retaining their distinguishability and taking into account only the relations to one another in which they are placed by the order-setting transformation ϕ, then are these elements called *natural numbers* or *ordinal numbers* or simply *numbers*, and the base-element 1 is called the *base-number* of the *number-series N*. With reference to this freeing the elements from every other content (abstraction) we are justified in calling numbers a free creation of the human mind. The relations or laws which are derived entirely from the conditions α, β, γ, δ in (71) and therefore are always the same in all ordered simply infinite systems, whatever names may happen to be given to the individual elements (compare 134), form the first object of the *science of numbers* or *arithmetic*.

§134 contains the proof of the categoricity of (second-order) arithmetic establishing that all systems (models) satisfying $\alpha - \delta$ are isomorphic. Arithmetic can thus disregard the particular nature of the objects composing its models and be concerned only with what follows from those conditions.

In the *Arithmetices Principia* (Peano 1889) and several editions of the *Formulaire de Mathématiques* (Peano 1895), Peano sets up to present arithmetic through a "new method," which first consists in presenting a background logical language and notation (which will inspire the formal setting of *Principia Mathematica*). Peano emphasizes that arithmetical "signs" are divided into those that can be defined from other arithmetical signs plus logical ones, and those that cannot. Their properties are characterized axiomatically (Peano 1889, pp. 85–86):

> If, as I think, these [signs] cannot be reduced any further, it is not pos-sible to define the ideas expressed by them through ideas assumed to be known previously. Propositions that are deduced from others by the oper-ations of logic are *theorems*; propositions that are not thus deduced I have called *axioms*. There are nine of these axioms (§1), and they express the fundamental properties of the signs that lack definition.

Primitive arithmetical notions are *number* (N), *unity* (1), *the successor of* $(a + 1)$, and *is equal to* (=). Peano takes numerical identity as a nonlogi-cal notion to be axiomatically defined, rather than taking identity as a prior logical notion that can then be applied to numbers (as for instance Frege did; Frege 1884, §63): as a consequence, four of his axioms are today seen as extra-arithmetical claims concerning identity. The other five, given (as Dedekind's conditions) within a theory of classes, are the proper axioms of arithmetic (presented here in modified notation):

PEANO'S AXIOMS

1. $1 \in N$
   - $a \in N \rightarrow a = a$
   - $a, b \in N \rightarrow (a = b \leftrightarrow b = a)$
   - $a, b, c \in N \rightarrow ((a = b \wedge b = c) \rightarrow a = c)$
   - $(a = b \wedge b \in N) \rightarrow a \in N$
2. $a \in N \rightarrow a + 1 \in N$
3. $a, b \in N \rightarrow (a = b \leftrightarrow a + 1 = b + 1)$
4. $a \in N \rightarrow a + 1 \neq 1$
5. $k \in K \wedge 1 \in k \wedge \forall x((x \in N \wedge x \in k) \rightarrow (x + 1 \in k)) \rightarrow N \in k$

Axiom 1 states that 1 is a natural number. Axiom 2 states that if a is a natural number, so is its successor. Axiom 3 states that the successor function is injective (different natural numbers have different successors). Axiom 4 states that 1 is not the successor of any natural number. Axiom 5 is the principle of mathematical induction – if k is a class, 1 is in k, and if x is a number in k then also $x + 1$ is in k, then the whole class N is in k. As regards Dedekind's conditions, α corresponds to Axiom 2; β entails both Axiom 5 – for the notion of chain is so defined as to encode mathematical induction – and Axiom 1 – since Dedekind's base-element 1 is contained in N by the definition of chain; γ corresponds to Axiom 4; and δ corresponds to Axiom 3.

As mentioned earlier (Section 3.1), and as is witnessed by the different conceptions of identity just rehearsed, the interplay between logical and nonlogical resources can significantly affect mathematical definitions. Standard modern presentations of the so-called *Dedekind–Peano Axioms* for arithmetic are given in second-order logic, and take 0 as the initial number:

PA^2 AXIOMS

1. $N(0)$
2. $\forall x \exists y (s(x) = y)$
3. $\forall x \forall y (s(x) = s(y) \rightarrow x = y)$
4. $\forall x (s(x) \neq 0)$
5. $\forall P (P(0) \wedge \forall x (P(x) \rightarrow P(s(x))) \rightarrow \forall x P x)$

Second-order arithmetic is categorical. Notice that the Induction Axiom [5] involves quantification over predicates. Since on a standard (Tarskian, referential) semantics second-order quantification over predicates is tantamount to quantification over sets (the sets of individuals to which the corresponding predicates apply), Quine (1970), §5, argued that second-order logic is not logic, but rather set-theory in disguise ("in sheep's clothing"): it is a mathematical theory itself and shouldn't be used in the foundations of mathematics. Not everyone shares Quine's concerns (Shapiro 1991), but this is still perceived as a serious challenge. Axiom 5 can also be formulated at first-order, avoiding this threat. In this case, however, further axioms need to be added, for addition and multiplication (and hence subtraction and division) can be defined from the PA^2 Axioms, but must be characterized axiomatically in first-order PA.

ADDITIONAL FIRST-ORDER PA AXIOMS

6. $\forall x(x + 0 = x)$
7. $\forall x\forall y(x + Sy = S(x + y))$
8. $\forall x(x \times 0 = 0)$
9. $\forall x\forall y(x \times Sy = (x \times y) + x)$

Most importantly, first-order arithmetic (due to the compactness of first-order languages) is not categorical, and has nonstandard models (i.e. models not isomorphic to the natural numbers; Boolos et al. 2007, ch. 25), hence the possibility of capturing a unique domain of objects (at least up to isomorphism) is jeopardized. Moreover, Axiom 5 becomes an axiom *schema* (quantifiers are dropped and predicate letters occur as schematic letters), expressing that any of its instances obtained by replacing schematic letters with suitable arithmetical open formulae is an axiom:

PA INDUCTION SCHEMA

5. $[\phi(0) \wedge \forall x(\phi(x) \rightarrow \phi(Sx))] \rightarrow \forall x\phi(x)$

   where x is free in any well-formed open formula $\phi(x)$.

Not only, however, one must clarify what knowing a schema amounts to (McGee 1997); also, Axiom 5 now stands for an infinity of sentences: first-order arithmetical axioms are not five, but infinitely many, and it seems to be a reasonable requirement on knowing a theory that it is finitely axiomatizable, so that all of its axioms can be surveyed at once. On the other hand, first-order logic is complete, while second-order logic isn't, thus making first-order logic a more secure setting to be working with. For these and other reasons, also weaker arithmetical systems are studied (the most relevant being PRA, Primitive Recursive Arithmetic, and Q, Robinson Arithmetic). However, PA and PA^2 remain standard reference axiomatizations in the foundations of mathematics.

Hilbert, Dedekind, and Peano (and his school; Burali-Forti 1894 ch. IV, §6; Padoa 1900; Cantù and Luciano 2021) all contributed to the canonization of the axiomatic method as a standard of mathematical theorizing (soon to be applied, for instance, to presentations of set-theories in the early nineteenth century). They also shared the intuition that axioms *jointly define* mathematical primitives (and, with Frege and Russell, sought to understand indefinables in *logic* too; Picardi 2022).

5.1.2 Roles and Criteria

CONSISTENCY is obviously a crucial requirement of axiomatic definitions (although research programs on inconsistent mathematics are also pursued; Weber 2022). When it comes to arithmetic, however, a dedicated discussion is needed. Hilbert's method to prove the consistency of geometry provided a relative consistency proof: Euclidean geometry is consistent if analysis (i.e. the theory whose objects are used to build the model) also is. The consistency of analysis remained an open question, as did the consistency of arithmetic itself. A relative consistency proof for arithmetic could be given only in logical terms, but the limitations of logicist programs made this hard to obtain. Hilbert hoped to find a direct consistency proof instead. He did this by taking the finitary part of arithmetic to be concerned with signs (syntactic strings) and by claiming that on its basis the whole of arithmetic, once suitably formalized in a system like that of Whitehead and Russell's *Principia Mathematica*, could be shown unable to syntactically derive both a formula and its negation (Hilbert 1926). Gödel's incompleteness theorems in 1930 and 1931 proved this to be impossible (second theorem), while also showing (first theorem) that any formal system capable of recovering enough arithmetic is *incomplete* (not every true consequence of the system is syntactically derivable in it). A later proof of the consistency of first-order PA was given by Gentzen, which nonetheless used methods (transfinite induction) that are not formalizable in PA itself. The debate on these issues is vast and cannot be rehearsed here (for surveys, see Raatikainen 2022; Zach 2023). Notice, however, that when it comes to the consistency of mathematical axiom systems, often only relative proofs can be obtained, based on other theories whose consistency is not directly proved, but rather taken as plausible enough given the extended time they have been used without engendering contradictions. Also, notice that while we are focusing here on criteria for axioms as definitions, other criteria (such as indeed completeness, soundness, etc.) can be imposed on the metamathematical properties of axiom systems more generally.

As regards other formal criteria, neither ELIMINATIVITY nor NON-CREATIVITY (CONSERVATIVENESS) meaningfully apply. For explicit (or implicit) definitions, these criteria guarantee that a newly defined symbol α of the extended language $\mathcal{L}^+$ can always be replaced by the *definiens* symbols $\beta_0, \ldots, \beta_n$ of $\mathcal{L}$, or that $\mathcal{L}^+$-sentences containing α can be replaced by $\mathcal{L}$-sentences containing only $\beta_0, \ldots, \beta_n$. But in axiomatic definitions the *definienda* are the primitives of $\mathcal{L}$: their elimination is tantamount to annihilating the theory altogether.

Axiomatic definitions have been traditionally considered implicit definitions proper (originally by Enriques; Biagioli 2023). An association between the two may have resided in an analogy with the principle of duality which Gergonne himself introduced in projective geometry, since the intersubstitutability of 'plane' and 'line' makes the individuation of their meaning a relational rather than absolute matter, close in spirit to how meaning is supposedly assigned to primitives by axioms (see also Pasch 1882 and Pollard 2010, Ch. 4). But the association is inadequate (Otero 1970; Gabriel 1978). First, they forfeit (G_3) (NUMERICAL BALANCE): generally each undefined symbol occurs in more than one axiom (hence they also fail Fregean SIMPLICITY). Also, they may fail to satisfy (G_4), UNIVOCALITY. As foreshadowed by Dedekind and displayed in Hilbert's independence and consistency proofs, axioms constrain, but do not necessarily fix univocally, the interpretation of the primitives. They only determine a set of relations, leaving room for indeterminacy as to what in fact their relata are. Any way of specifying their meaning will thus prioritize structural features (Giovannini and Schiemer 2019 aptly call them "structural definitions"), and for this reason many proposals on this score underlie different varieties of mathematical structuralism.

Starting from Dedekind's work and later suggestions by Benacerraf (see Section 3.2), structuralism has become one of the most developed philosophical accounts of arithmetic and mathematics generally (Hellman and Shapiro 2018). One of its major methodological strengths is that it conjoins philosophical inquiry with a focus on those structural relations which to many seem to matter to professional mathematicians much more than preoccupations with the nature of mathematical objects. Structuralism thus seems especially adequate as a philosophical account of actual mathematics. Also, it is closer in spirit to some algebraic traditions in mathematics (think of algebra, or the views expounded by Bourbaki; Bourbaki 1950), as well as to more recent discussions on the foundational role of category theory (Reck and Schiemer 2023, §3).

For our present purposes (although comparison between structuralism and rival views will resurface below), we are just interested in outlining how axiomatic or structural definitions can provide knowledge of their subject matter, hence how their semantic role can be differently understood, either by reinstating UNIVOCALITY somehow, or by motivating how to dispense with it. Here are some possible options:

STRUCTURAL OBJECTS. Axioms define univocally one specific object, a *structure* (progressions, complete ordered fields, etc.). This has been advocated by Shapiro's *ante rem* structuralism, which is eliminative about individual mathematical objects (unless we conceive of them as places in structures in what Shapiro calls a places-as-objects perspective), but noneliminative about structural objects (Shapiro 1997). Structures are conceived analogously to universals. They are self-subsistent ("free-standing," Shapiro 1997, 92–96) abstract objects, which exist whether they are instantiated or not (hence, *ante rem*). This yields a peculiar kind of platonism about structures. As such, *ante rem* structuralists must provide identity conditions for structures, and account for our epistemic access to them (Shapiro 2011). In other variants (Hellman and Shapiro 2018), structures are conceived as patterns or systems of relations that particular (actual or possible) systems of objects instantiate, their existence being dependent on the (actual or possible) existence of such systems (notice also that proposals to define structures via abstractive definitions have been advanced: Linnebo and Pettigrew 2014; Leach-Krouse 2017).

PARADIGMATIC MODELS. Axioms do capture a specific *intended* model. For arithmetic, for instance, this can be provided by the natural numbers as *sui generis* objects, or by a particular model of sets (usually the von Neumann ordinals we already mentioned in Section 3.2.5) in set-theoretical varieties of structuralism. Pending a reply to Benacerraf's challenge, UNIVOCALITY is defended, and any other model satisfying a certain structure will count as an "isomorphic imposter" (Shapiro 2000, 361) – although (categorical) axioms will guarantee that any structure-preserving translation of the theory will hold of any such isomorphic model. Methodologically, one can accept that different isomorphic models for the same axiomatic definition are available, and pick a preferred one (e.g. a particular set-theoretical or category-theoretical construction) for theoretical reasons such as pervasiveness in mathematics, perspicuity, manageability, and so on. Such *relativist structuralism* (Reck and Price 2000) sidesteps the challenge of multiple realizability, while at the same time avoiding the need to posit the existence of structures.

RELATIONAL OBJECTS. Axioms define first-level concepts of objects, but these are entirely individuated by *relational* properties. This reinstates UNIVOCALITY, but has the metaphysical burden of explaining how objects can only have relational properties (and, prior to that, how structural properties should be characterized: Korbmacher and Schiemer 2018).

ARBITRARY REFERENCE. Axioms bestow objectual reference on primitives but reference is understood as *arbitrary*: for instance, '0' refers arbitrarily to *any* object which may occupy the first position of an ω-sequence, although not to any specific one (Boccuni and Woods 2020). This clearly requires defending arbitrary reference as tenable (see e.g. Breckenridge and Magidor 2012).

PARTIAL AND COLLECTIVE DENOTATION. Axioms define primitives univocally, but semantic reference, although not being arbitrary, is partial: '0' partially denotes *all* objects occupying the first position in all (actual or possible) systems instantiating an ω-sequence. Reference is not univocal, but still not indeterminate (against Quine's views; see Field 1974, 220–223). Alternatively, one can take numerals to refer collectively to all objects occupying a given position in all systems instantiating a given structure (White 1974).

CONCEPT STRUCTURALISM. Axioms primarily define concepts, and whether these are concepts of specific individual objects, and which ones, is either impossible to establish, or immaterial. As for the first option, Frege, in his exchange with Hilbert on geometry (Blanchette 2018), objected that Hilbert's axioms define not first- but *second- (or higher-) level concepts*. They do not establish, for example, what objects natural numbers or geometrical points are, but rather what it takes for a concept to be a natural-number-concept or a geometrical-point-concept:

> The characteristic marks you give in your axioms are apparently all higher than first-level; i.e., they do not answer to the question 'What properties must an object have in order to be a point (a line, plane, etc.)?', but they contain, e.g., second-level relations, e.g., between the concept point and the concept line. (Frege 1980, Frege to Hilbert, 6.1.1900, 46; also Frege 1971, §II)

Hilbert somehow concurred (similar remarks are found in Pasch 1882; see Hodges 2023):

> You say that my concepts, e.g. 'point', 'between', are not unequivocally fixed; e.g. […] point is […] a pair of numbers. But it is surely obvious that every theory is only a scaffolding or schema of concepts together with their necessary relations to one another, and that the basic elements can be thought of in any way one likes. (Frege 1980, Hilbert to Frege, 29.12.1899, 40)

If we follow Frege's criticism, axioms merely define conditions for certain concepts to be concepts of kind F, but do not allow establishing univocally which objects F's are. Bracketing Frege's criticism, however, this view can lend support to a positive proposal where concepts take explanatory priority over objects, namely to a form of *conceptual structuralism* (Reck and Schiemer 2023, §2.2; Ferreirós 2023) according to which mathematics deals primarily with relations among concepts, rather than with objects or structures conceived as such.

The axiomatic method has now a long tradition, and is pervasive in mathematical practice. For this reason, an exhaustive discussion of its theoretical and philosophical roles would need to consider a vast array of purposes, including its ability to provide rigorous analysis of mathematical notions, of permitting suitable architectonic systematizations of different parts of mathematics, of serving heuristic or even pedagogical functions (Cantù 2023). Here we'll confine ourselves to some brief remarks on those roles that we have explored in connection with explicit and implicit definitions in general.

Salience of the *definienda* is straightforward: primitives are necessarily salient within a theory. Salience of the *definiens*, namely the axiom system, boils down to the motivation for having the theory in the first place.

Ontological Reduction may be accomplished in a peculiar sense. We could first define reals (e.g. following Cantor or Dedekind) as set-theoretical constructions of rationals, and then provide axioms for real number theory. However, these axioms would not introduce reals as primitives, but simply point to objects independently defined: reduction is accomplished beforehand. If, on the contrary, we outright lay down the axioms, no reduction is at stake. An indirect form of reduction is afforded when an axiomatic theory M offers a reduction basis for another theory M^*, and the primitives of M^* are in the end reduced to the objects axiomatically defined in M (as is the case for most mathematical theories with respect to set theory or category theory).

To some extent, axiomatic definitions can provide both **Conceptual Analysis** and **Rational Reconstruction** (as well as being the outcome of a process of Explication; see Section 6.2). They can systematize informal notions inherited from practice, or provide a way to reconstruct our knowledge in a certain domain. For a familiar example among the many, Hilbert's claim that the choice of the axioms of Euclidean geometry "is tantamount to the logical analysis of our intuition of space" could be seen in this light. Frege emphasized that axioms can be ambiguous between analyses of antecedent notions and arbitrary stipulations (Frege 1980; Frege 1971, §3). Both eventually took the connection between informal notions and formal rendition as more inspirational than constitutive: once a formal system has been laid out, the connection between the informal notion and the formalized one can be disregarded (Section 7.3.2).

5.2 Definitions by Abstraction

5.2.1 Outline

Definitions by abstraction are discussed in the foundations of mathematics at least since Frege's works (and adopted both in Frege's and earlier times; Mancosu 2016), and still animate a vast debate (Boccuni and Zanetti in press). They are provided by *abstraction principles* of the form:

$$\forall\alpha\forall\beta\ [\Sigma(\alpha) = \Sigma(\beta) \leftrightarrow \alpha \sim \beta] \tag{ABS}$$

Syntactically, 'Σ' is a term-forming operator taking expressions like 'α' and 'β' as arguments and delivering singular terms as values, and '$\sim$' is a relation-term of the underlying language. Semantically, the abstraction function Σ takes the denotata of 'α' and 'β' as arguments and delivers the denotata of '$\Sigma(\alpha)$' and '$\Sigma(\beta)$' as values, and '$\sim$' stands for an equivalence relation between α's and β's. The principle can then be instantiated by different choices of 'α', 'β', and '$\sim$' in its right-hand side (RHS).

Frege appealed to an instance of ABS in attempting a definition of the concept of (finite) cardinal number. In Frege (1884), after surveying a vast array of rival views, some cornerstone claims are offered:

CONCEPT-ASCRIPTIONS: numerical ascriptions – *Zahlangaben*, namely statements of the form 'Mars has two moons' – are statements *about concepts*. They state that the relevant concept (e.g. ⌜Moons of Mars⌝) is such that a particular number is associated to it.

SINGULAR TERMHOOD: the proper logical form of numerical ascriptions is that of identity statements ('The number of the concept ⌜Moons of Mars⌝ = 2'), with numerical expressions occurring as *singular terms* in substantival position, hence purporting to refer to individual, "self-subsistent" objects.

CARDINALITY: expressions of the form 'The number of the concept F' stand for *cardinal numbers* (*Anzahlen*), and answer 'How many?'-questions about the objects falling under a concept F.

CONTEXT PRINCIPLE: the meaning of words should never be sought for in isolation, but always in the context of a sentence (Frege 1884, Introduction and §62) (ruling out, e.g. psychologistic conceptions of linguistic meanings as mental representations).

It follows that the (inherently epistemological) question "How are numbers given to us?" (Frege 1884, §62) can only be answered by establishing the meaning of numerical identities. Frege's insight is that "whenever we speak of objects of any kind, we must have in the background a principle for determining what is to count as the *same* object of that kind" (Dummett 1991, p. 162). Fixing the meaning of identity statements between numerical terms gives us a criterion of identity for the objects they refer to, insofar as they "express our recognition of a number as the same again" (Frege 1884, §62) even when referred to by means of different terms. Such numerical identities should thus be part of the very definition of cardinal number (*Anzahl*), which Frege (inspired by a passage by Hume) first attempts through what is known today as *Hume's Principle*:

$$\forall F \forall G \, (\#F = \#G \leftrightarrow F \approx G). \tag{HP}$$

Hume's Principle states that for any two concepts F and G, the number of F is identical to the number of G *iff* F and G are equinumerous, that is, can be put into a one-to-one correspondence relation (bijection).

Hume's Principle is a second-order abstraction, with terms for second-order entities (concepts) in the RHS, and terms for first-order entities (objects) in the LHS. A first-order abstraction is the Direction Principle (DP) also discussed by Frege:

$$\forall a \forall b \, (D(a) = D(b) \leftrightarrow a//b) \tag{DP}$$

stating that, for any line a and b, the Direction of a is identical to the Direction of b *iff* a and b are parallel.

Hume's Principle and DP contextually define their *definienda* ('$\#F$', '$D(a)$') by means of the sentential conditions on their RHS: Frege suggests (Frege 1884, §64) that we "carve up the content" of the RHS "in a way different from the original one, and this yields us a new concept." How recarving, or reconceptualization, is best understood is controversial (Hale 1997; Linnebo 2018b, ch. 2). According to Frege, we "remov[e] what is specific in the content" of the equivalence in the RHS, and we "divid[e] it" between the two related items.

Such *logical abstraction* turns an equivalence relation into an identity between objects – which are objects *of* α and β (as numbers, for Frege, are numbers of concepts).

Hume's Principle has significant definitional potential. Its *definiens* (the RHS) can be formulated logically – in a second-order logic, defining equinumerosity requires quantifying over relations:

$$\exists R \, [\forall x(Fx \rightarrow \exists!y(Gy \wedge R(x,y)) \wedge \forall x(Gx \rightarrow \exists!y(Fy \wedge R(y,x))))] \qquad (\approx)$$

An initial true instance of HP is obtained by instantiating the RHS with a purely logical concept, $\ulcorner$being non-self-identical$\urcorner$, $\ulcorner x \neq x \urcorner$ – where identity, following Leibniz, can be second-order defined by saying that objects are identical *iff* they share the same properties: $\forall x \forall y(x = y \leftrightarrow \forall F(Fx \leftrightarrow Fy))$. This first instantiation then allows the definition of infinitely many cardinals through a procedure known as Fregean Bootstrapping:

> ### FREGEAN BOOTSTRAPPING
> Take the logical concept $\ulcorner x \neq x \urcorner$, and substitute it for both variables F and G in the RHS of HP. Since equinumerosity is reflexive, this instance of the RHS of HP is true. It follows by stipulation that the corresponding LHS is true, hence that $\#(\ulcorner x \neq x \urcorner) = \#(\ulcorner x \neq x \urcorner)$. From here, apply Existential Generalization to derive that $\exists x(x = \#(\ulcorner x \neq x \urcorner))$. This first-order existential quantification shows (under most interpretations) that there exists an object which is the number of the concept $\ulcorner$being non-self-identical$\urcorner$. Define then explicitly: $0 =_{df} \#(\ulcorner x \neq x \urcorner)$. Use 0 to define a new concept, namely $\ulcorner x = 0 \urcorner$. Apply the same procedure as earlier, and obtain the existence of $\#(\ulcorner x = 0 \urcorner)$, which explicitly defines '1'. Use 1 to define the concept $\ulcorner x = 0 \vee x = 1 \urcorner$ hence deriving the existence of $\#(\ulcorner x = 0 \vee x = 1 \urcorner)$, which explicitly defines '2'. Proceed similarly for subsequent cardinals.

Further definitions deliver the notion of finite cardinal, i.e. natural, number. Frege (1879), III, 76, defines the notion of *following in a R-series*, or *ancestral* of a relation R. Take a relation R. Define a concept F as *hereditary* in the R-series *iff* for any two objects x and y, if $R(x,y)$, then if $F(x)$, then $F(y)$: F is inherited in R-series from any x which is F by any y with which x bears R. The ancestral of R, R^+, is defined by saying that y follows x in the R-series *iff* y falls under all those R-hereditary concepts under which x and any object z such that $R(x,z)$ fall. Compare $\ulcorner x$ is the father of $y \urcorner$ (R) and $\ulcorner x$ is the ancestor of $y \urcorner$ (R^+). *Weak ancestral*, R^-, is then defined as: y follows x in the R-series or $y = x$. Compare

⌜x is the ancestor of y⌝ (R^+) with ⌜x is a member of the lineage of y⌝ (R^-), which reflexively applies to the originator of the lineage. Frege then logically defines the relation S of *immediate successor* between cardinals m and n, and can thus define the (ancestral) relation $S^+(n, m)$ ⌜being greater than⌝ ($m < n$), and (weak ancestral) $S^-(n, m)$ as ⌜being greater than or equal⌝ ($m \leq n$). With $m = 0$, this provides the relation of being greater or equal to 0, that is, following in the S-series beginning with 0 or being equal to 0. This condition defines the concept of finite cardinal (i.e. natural) number. It does so via logical principles and definitions only. This would qualify arithmetical knowledge, and epistemic access to numbers, as logical and *a priori*. It also would show (a) that some objects are given to us through purely intellectual processes without appeal to intuition (*contra* Kant) or mental representations (*contra* empiricists); (b) that arithmetical statements are analytic in Frege's sense of being provable by logic and definitions (more on this in Section 5.2.2).

HP provides a simple procedure to establish whether two numerical terms refer to the same cardinal (hence providing conditions of identity). We extract the concepts F and G from '$\#F$' and '$\#G$' and check whether they are equinumerous. If they are (the RHS is true, hence the LHS is true, hence), they refer to the same cardinal; if they aren't (the RHS is false, hence the LHS is false, hence), they refer to different cardinals. Given a well-defined F, however, Frege's COMPLETENESS criterion also requires us to be able, for any x, to establish whether x is or is not an F. But take any sentence of the following form:

$$\#F = q, \tag{CS}$$

where 'q' is a singular term not of the form '$\#F$' (nor previously introduced as abbreviating such a term), such as 'England' or 'Julius Caesar.' There is no G to be extracted from q and be compared with F as regards equinumerosity. HP does not establish, for any x, whether x is a cardinal (hence fails to provide conditions of application). This is known as the *Caesar Problem* (CP) and is a symptom that HP fails ELIMINATIVITY, for it cannot eliminate its *definiendum* in all sentential contexts, but only in those of the form '$\#F = \#G$' (see also Section 5.2.2 below).

Frege then discards HP as a definition, and quite surprisingly reverts to an explicit definition. He avails himself of what he initially considers a logical notion, that of *extension* of a concept, this being an object containing the elements falling under a concept. One way to understand it is as a proto-set-theoretical notion, roughly coinciding with a naive notion of set. Frege (1884), §69, defines '$\#F$' explicitly as 'the extension of the concept ⌜being equinumerous to the concept F⌝'. If we let '$\varepsilon(F)$' stand for 'the extension of the concept F' and ⌜$\approx F$⌝ be the second-level concept of concepts ⌜being equinumerous to the concept F⌝, Frege's explicit definition is:

$$\#F =_{df} \varepsilon(\ulcorner \approx F \urcorner). \tag{$N^=$}$$

This identifies cardinal numbers with equivalence classes of equinumerous concepts, providing a specific kind of logicist platonism about cardinal numbers (the definition in Frege 1893–1903 differs slightly and defines cardinals as classes of extensions of equinumerous concepts). Being explicit, $N^=$ is eliminative: every arithmetical statement becomes a notational variant of a logical statement, and both identity and application conditions are guaranteed, provided we know what extensions are.

Extensions get formal treatment in Frege (1893–1903). Again, Frege appeals to an abstraction principle. His *Basic Law V* introduces the crucial notion Frege works with in *Grundgesetze*, that of *value-ranges* of functions, stating that two functions have the same value-range *iff* they have the same values for the same arguments. When applied to those particular functions (from objects to truth-values) that concepts are for Frege, this provides extensions:

$$\forall F \forall G (\epsilon(F) = \epsilon(G) \leftrightarrow \forall x(Fx \leftrightarrow Gx)). \tag{BLV}$$

Any concepts F and G have the same extension *iff* the same objects fall under both (notice that for reasons internal to the system of *Grundgesetze* and related to the so-called *proof of referentiality*, an equivalent CP for extensions may be avoided; (see Frege 1893–1903, §10; Heck 1999; Linnebo 2004; Bentzen 2019).

Frege shows that all arithmetical truths (including, most notably for us, the PA^2 Axioms) can be derived as *theorems* from BLV via $N^=$ (and HP, which can itself be derived as a theorem – but see May and Wehmeier 2019). In his setting we have no *proper axioms* of arithmetic, that is, unprovable arithmetical first principles: the only basic laws are (supposedly) logical. The Induction Axiom itself is a consequence of the definition of cardinals. If BLV is taken as a definition of extensions, Frege's *logicism* can be seen as the view that arithmetical statements are derivable from logical laws as notational variants of logical statements obtained via logical definitions (on varieties of logicism, see Boccuni and Sereni 2021).

Unfortunately, the introduction of BLV in the system of *Grundgesetze* (Cook 2023) makes the latter inconsistent, as discovered by Russell (Frege 1980, letter 16.06.1902). Extension bears strong resemblance with a naive notion of set (Burge 1984), and Russell's Paradox affects both (hence also Cantor's, Dedekind's, and Peano's settings). Two principles underly the latter:

PRINCIPLE of EXTENSIONALITY (PE): $\forall x \forall y (\forall z(z \in x \leftrightarrow z \in y) \rightarrow x = y)$: sets with the same elements are identical.

UNRESTRICTED COMPREHENSION PRINCIPLE FOR SETS (UC): $\exists x \forall y(y \in x \leftrightarrow \phi(y))$, where ϕ is any formula not containing y free: for any open formula ϕ not containing y free there exists the set $\{x : \phi(x)\}$ of the objects satisfying ϕ.

Take the property $\mathcal{R} = x \notin x$ ('x is not a member of itself'). By UC, the set $R = \{x : x \notin x\}$ exists. Since sets are objects in the first-order domain, R is among the possible values of 'x' in $\mathcal{R}$. So one can ask whether $R \in R$ or $R \notin R$. Suppose $R \in R$; then R must satisfy the condition for membership in R, namely $\mathcal{R}$; hence $R \notin R$. Suppose $R \notin R$; then R satisfies condition $\mathcal{R}$; thus $R \in R$. Hence the contradiction – and the failure to satisfy PE, since we cannot know which elements are in R and thus whether R is identical to itself. In Fregean terms, this can be formulated by taking the concept ⌜x is the extension of a concept F, and x does not fall under F⌝, then by taking the extension K of this concept, to see that K is an element of K if and only if it isn't (Frege 1893–1903, II, Afterword, 254).

Why does the contradiction arise? Intuitively, BLV makes incompatible requests. Its right-to-left direction (Va) imposes a functional correlation between concepts and extensions: different extensions are associated to different concepts. But the opposite direction (Vb) entails that such correlation is one-one: distinct concepts are correlated with different extensions. Since UC entails that there exists a concept under which all and only the objects satisfying the RHS of BLV fall, we have a conflict: by UC there must be more concepts than extensions, while by Vb there must be as many extensions as concepts. If concepts are extensionally identified with subsets of the first-order domain of objects, this seems to engender paradox by violating Cantor's Theorem, according to which the cardinality of the power set $\mathcal{P}(x)$ (i.e. the set of all subsets) of a set x with cardinality k is 2^k and hence strictly greater than x.

However, the inconsistency may reside in other interactions with the logical system of GGA (Zalta 2023; Cook 2023; Boccuni and Sereni in press). Some (Dummett 1993a) have blamed UC (or, better, the principles in Frege's system – like his Rule of Substitution – which play an equivalent role), especially for its being impredicative. *Impredicativity* can apply to definitions themselves, or to their background logic. A definition is impredicative if the *definiendum* is defined via reference to a domain of objects to which it itself belongs. Basic Law V is impredicative in this sense: it appeals to a first-order domain (the one of its RHS) which itself contains extensions as first-order objects. As regards the background logic, UC is impredicative because ϕ is allowed to contain bound second-order variables. Since it is by the impredicativity of UC that $\mathcal{R}$ can be formed, restricting UC predicatively may prevent the paradox. The drawback is that the system of GGA significantly loses mathematical strength (see e.g. Burgess 2005). Analogous restrictions, applied directly to BLV, may have similar consequences. In recent times, several ways of adjusting BLV so as to make it consistent have been advanced (starting from Boolos' New V; see Boolos 1986), also as ways of reinstating Frege's program for set-theory

(Studd 2016; Linnebo 2018b, ch. 12; Cook 2023). Here we must emphasize that the culprit of the contradiction cannot just lie in the logical form of BLV, since other abstraction principles, including HP, are consistent.

As anticipated, from BLV and $N^=$, HP can be derived as a theorem rather than being introduced as a definition. *ELIMINATIVITY* isn't thus here a constraint for HP, but is *a fortiori* satisfied since any occurrence of '$\#F$' in HP can be eliminated via $N^=$. The status of abstraction principles as primitive implicit definitions, then, is not just given by their form, but by their theoretical role within a theory.

The possibility of conceiving of HP itself as a definition has been vindicated by (Scottish) Neologicism (Wright 1983; Hale and Wright 2001a; Cook 2007; Ebert and Rossberg 2016). Thanks to formal results by Boolos (1987) and Heck (1993), it has been shown that (a) in the *Grundgesetze* the appeal to extensions is indispensable up to the derivation of HP and dispensable thereafter, and (b) the system known as *Frege Arithmetic*, that is, the system of impredicative second order logic with the addition of HP as a definitional axiom (*i*) is consistent (if PA^2 is), and (*ii*) proves the PA^2 Axioms, via a derivation known as *Frege's Theorem* (Heck 2011; Zalta 2023), essentially outlined earlier. The (envisioned) significance of BLV, and the contemporary significance of HP, are best clarified by surveying their expected roles and criteria.

5.2.2 Roles and Criteria

Despite similarities, it will be useful to discuss the Fregean and neologicist settings separately.

Fregean Definitions

Bracketing the inconsistency of BLV, the explicit definition $N^=$ satisfies NONCIRCULARITY, ELIMINATIVITY, SUBSTITUTABILITY SALVA VERITATE, SYNONYMITY, as well as COMPLETENESS and SIMPLICITY. It satisfies FRUITFULNESS through its role in providing proofs of arithmetical theorems. It would satisfy CONSISTENCY and NONCREATIVITY (CONSERVATIVENESS) if the background theory were consistent.

Salience (of both *definiens* and *definiendum*) is obviously an intended role. Frege's analysis would show that cardinal numbers are extensions, and which extensions they are. In a sense, $N^=$ also provides **Ontological Reduction**: cardinals are just extensions.

Regarding **Conceptual Analysis**, things are more complicated. Frege's aim both in *Begriffsschrift* and the *Grundlagen* can reasonably be seen as an attempt at capturing the notion of cardinal and natural number inherited from mathematical practice. But in *Grundgesetze* and later writings (Frege 1979a)

Frege suggests that once a proper formal analysis is completed, the original informal notion should be abandoned. As anticipated, his attitude converges with Hilbert's: analysis has more of an heuristic role, and adequacy to actual practice is more an *ex post* virtue of the formal system in which the notion is characterized than a strict *ex ante* criterion for the construction of the system (for discussions, see e.g. Antonelli and May 2000; Blanchette 2007; Shieh 2008; Hallett 2021).

Basic Law V can be seen – though strikingly Frege (1893–1903), II, §146, himself disagrees – as a primitive implicit definition of a logical notion. As such, it should satisfy a criterion of LOGICALITY. It is hard to pin down what a logical definition consists in, however (a criterion for the logicality of an expression, inspired by Tarski, that can be applied to abstractions too, is permutation invariance: see Antonelli 1998). At least, a logical *definiendum* should be introduced via a logical *definiens*. Bracketing inconsistency, BLV accomplishes this. But much depends on what counts as logical in the first place. A feature traditionally credited to logic is topic-neutrality (logical claims should hold in any domain of discourse), which entails that logic should not be concerned with particular objects. This, however, is clearly alien to Frege's conception of logic, where extensions are logical objects. Other features sometimes credited to basic logical principles are self-evidence or obviousness. But Frege himself, after the paradox, retrospectively claims that BLV "is not as obvious as the [other basic laws] nor as obvious as must properly be required of a logical law" (Frege 1893–1903, II, Afterword, 253). Moreover, clarifying these notions may be as hard as clarifying the notion of logicality (Jeshion 2001; Jeshion 2004; Shapiro 2009).

If taken as a primitive implicit definition, BLV should fail ELIMINATIVITY and SUBSTITUTABILITY *salva veritate*, and it does: only occurrences of the *definiendum* which appear in contexts like those of the LHS can be eliminated. It obviously fails CONSISTENCY. It satisfies SIMPLICITY (but see again Frege 1893–1903, II, §146) and NONCIRCULARITY, and had it been consistent, would have satisfied COMPLETENESS and FRUITFULNESS. As any implicit definition, it fails term-to-term SYNONYMITY between *definiendum* and *definiens*, although some form of semantic equivalence will have to be preserved between the defining (RHS) and the defined (LHS) sentential contexts. As a primitive implicit definition, BLV should fail NONCREATIVITY (it should be nonconservative), but it ends up being so vacuously (in a classical setting), given its inconsistency.

Frege discusses noncreativity in a sense different from conservativeness, endorsing a criterion that we may call FREGEAN NONCREATIVITY, its major target being Dedekind (Frege 1893–1903, Vol II, §§138 ff.). Dedekind claimed that "numbers are free creation of the human mind" (Dedekind 1888), that for any nonrational cut "we create for ourselves a new, irrational number"

(Dedekind 1872), and that expansions of number systems (from $\mathbb{N}$ to $\mathbb{Z}$ to $\mathbb{Q}$ etc.) are "formed by a new creation" (Dedekind 1888). Frege takes this to mean that definitions are mental acts of metaphysical creation. He opposes this for his antipsychologism and for his firmly objectivist view of mathematics, according to which "Just as the geographer does not create a sea when he draws borderlines […] so too the mathematician cannot properly create anything by his definitions" (Frege 1893–1903, I, Preface, xiii). In a Fregean (as well as a neologicist) conception, abstraction does not create new objects, but merely provides new conceptual means to refer to, or think of, objects which were there all along.

Strikingly, he wonders about BLV itself whether "our procedure [can] be called a creation" (Frege 1893–1903, II, §147). This can end up in a "quarrel about words," the main point being that "our creation, if one wishes so to call it, is not unconstrained and arbitrary, but rather the way of proceeding, and its permissibility, is settled once and for all." What Frege most strongly opposes is that definitions can be used as a way for a free and unconstrained creation of mathematical objects. His major concern is to provide nonarbitrary principles for the adoption of definitions (Ebert and Rossberg 2019; Hallett 2019). Whether Dedekind deserves Frege's charges is an entirely different matter, given that his definitions of the irrationals and his axiomatization of arithmetic are certainly rigorous by mathematical standards, and have proved their worth in becoming entrenched parts of mathematics. It is also worth noting that some authors gave more constructivist readings of abstraction, seeing them in general as akin to creative definitions (e.g. Weyl 1949, §1.2, takes them as definitions "through which new ideal objects can be generated," akin to those of points at infinity or other ideal elements in geometry and mathematics).

A complete discussion of definitions in Frege cannot ignore two essential criteria, Sortality and Analyticity. Since they are crucial also for neologicists, we discuss them in the next section.

Neologicist Abstractions

For Frege's Theorem to have a philosophical significance, HP must be vindicated as a definition. Hence, a solution to the Caesar Problem must be provided. To appreciate the relevance of CP, a further criterion on Fregean definitions for *some* concepts must be noted:

SORTALITY. A definition introduces a sortal concept *F iff* it provides it with both conditions of identity and conditions of applications.

Sortality (Dummett 1991, ch. 13; Wright 1983, ch. 1) can be seen as a specification of the COMPLETENESS criterion that applies to concepts under which determinately distinct objects fall, hence to any concept *of* objects which isn't vague (a precondition, according to Frege, to be able to count the objects falling under the concept, and hence to apply arithmetic). It doesn't pertain to mass concepts (gold, water, ...) or to color concepts (red, ...) because they are not concepts under which objects fall. One possible diagnosis of CP, as hinted earlier, is that HP doesn't provide a sortal concept of cardinal since it doesn't determine conditions of application.

Rather than a formal solution, neologicists offer a philosophical one (Hale and Wright 2001b), by endorsing a particular "philosophical ontology": "all objects belong to one or another of a smallish range of very general categories, each of these subdividing into its own respective more or less general pure sorts; and in which all objects have an essential nature given by the most specific pure sort to which they belong" (389).

This picture is backed up by a Sortal Inclusion principle (SI), establishing that a sort of objects F is included under a sort G only if some suitable identity statements about G's are true because they satisfy the same identity criteria for identity among F's (Hale 1994, 198; see also Wright 1983, §xiv). Once applied to concepts of cardinal numbers and persons this principle (or a particular specification of it, called N^d) entails that cardinals can be persons only if some identities among persons are true because they satisfy the same identity criteria for identity among cardinals, namely because a certain one-one correlation obtains between certain concepts. Caesar can be a cardinal only if he falls under a sortal concept whose identity conditions are exhaustively provided in terms of equinumerosity among concepts. Since he doesn't (for identity conditions for persons are not of this kind), SI and N^d tell us that persons are not cardinals, and vice versa. Caesar is not a cardinal and the Caesar sentence is false. Since objects fall under many sortals, it is crucial to identify those that somehow capture their essence. Sortal Inclusion and N^d should thus be restricted to *pure sortals*, such that if an object is an instance of a pure sortal, then it is so necessarily and "could not survive ceasing to be so" (Hale and Wright 2001b, 387). Restriction to pure sortals is also required to deal with another problem, namely to guarantee that the RHS of HP exhibits everything that is involved in the essence of numbers, ruling out that numbers can have any *additional nature* that HP isn't able to capture (Hale and Wright 2008; Potter and Sullivan 2005; Hale and Wright 2008) – relatedly, one could treat HP as a *real* definition which completely captures the nature of cardinal numbers (Gideon Rosen 2020).

Even granting a solution to CP, neologicists must vindicate arithmetic as analytic, at least in Frege's sense. There is a wide debate on whether Frege's aims were preeminently philosophical or mathematical, and this affects the weight that should be put on analyticity as either, respectively, an essential philosophical desideratum in its own right, or rather a by-product of the formal reconstruction of arithmetic (thus relating to a broader debate on the combination of mathematical and philosophical concerns in Frege's views; Benacerraf 1981; Weiner 1984; Blanchette 1994; Panza and Sereni in press). Fregean analyticity may be connected to both, since it strictly relates to the notion of proof:

FREGE-ANALYTICITY. A statement is Frege-analytic *iff* in "finding the proof of the [statement], and […] following it up right back to the primitive truths … we come only on general logical laws and on definitions" (Frege 1884, §3).

The analytic/synthetic distinction dates back to Hume's distinction between *relations of ideas* and *matters of fact* (Hume 1748, IV, 1). The guiding thought is that some statements can be true just in virtue of the meaning of their component expressions, with no contribution from empirical evidence. Analyticity has been crucial in defending the possibility of *a priori* knowledge in rationalist views. Logical empiricism relied heavily on analyticity too, since it would guarantee truth by conventional stipulations (Section 7.2.1) to logical and mathematical statements which could not be empirically verified. Analyticity is clearly connected with definitions, as is synonymy: by conventionally stipulating a definition we establish the synonymy between *definiens* and *definiendum* and we secure its truth merely in virtue of meaning. At least with regards to the logical empiricist views, Quine (1951) challenged the tenability of the notion, primarily on two grounds: that analyticity and synonymy are essentially interdependent notions and cannot thus be used to explain one another, and that there is no clear way of distinguishing the contribution of either world or language in the meaning of a statement.

Despite this, the notion is still defended today. Following Boghossian (Boghossian 1996b), we can distinguish *Epistemic Analyticity* – "a statement is 'true by virtue of its meaning' provided that grasp of its meaning alone suffices for justified belief in its truth" (Boghossian 1996a, 334) – and *Metaphysical Analyticity* – "a statement is analytic provided that, in some appropriate sense, it owes its truth-value completely to its meaning, and not at all to 'the facts'" (*ibid.*). Neither exactly matches Frege Analyticity, although Epistemic Analyticity may come closer to what Frege had in mind, for the role of 'meaning' and 'justification' mirror those of 'definitions' and '(logical) proof' in Frege's notion. After all, Frege believed that a judgment on the analyticity of a statement "is a judgement about the ultimate ground upon which rests the justification for holding it to be true" (Frege 1884, §3).

Frege presents his notion as a clarification of Kant's notion of analyticity (Kant 1781, Introduction, IV), which was heavily influenced by a traditional

Aristotelian conception of logic and definitions, as concerned with relations of inclusion among concepts, and based on a logical analysis of judgments that mirrors the grammatical distinction between subject and predicate. Focusing on universal affirmative judgments ('All A's are B's'), analytic judgments are such that "the predicate B belongs to the subject A as something that is (covertly) contained in this concept A," and "the connection of the predicate is thought through identity," namely by the Law of Non-Contradiction (so that it is not possible to deny without contradiction that all A's are B's) – contrary to synthetic judgments where the predicate "B lies entirely outside the concept A." As a consequence, analytic judgments are "judgments of clarification [...] since through the predicate [they] do not add anything to the concept of the subject, but only break it up by means of analysis into its component concepts, which were already thought in it (though confusedly)": only synthetic judgments ("judgments of amplification") increase knowledge.

In contrast, Frege's notion of analyticity is epistemic (and proof-theoretic) in nature: the question of analyticity is "removed from the sphere of psychology," and as anticipated, concerns the justification for holding the content of a statement (*Satz*) true. The end purpose is to show, *contra* Kant, that analytic statements can be informative and increase our knowledge. It is impossible to understand the import of Frege's notion without considering the radical changes in its background logic and semantics (MacFarlane 2002; Linnebo 2003). These entail that logical form need not respect the grammatical subject/predicate distinction, being rather based on the function/argument distinction. This allows Frege to consider the analyticity of statements lying outside the scope of traditional logic: existential statements ($\exists x(Fx)$), simple predications ($F(a)$), and relational statements ($R(a, b)$). Analytical consequences of axioms and definitions increase our knowledge by entailing statements whose truth cannot be ascertained by a mere decompositional analysis of concepts: "they are contained in the definitions, but as plants are contained in their seeds, not as beams are contained in a house" (Frege 1884, §88).

Is neologicist arithmetic analytic? While arithmetical statements can't be just notational variants of logical ones (since HP is noneliminative), HP and Frege's Theorem seem to entail that PA^2 Axioms are *a priori* and Frege-analytic. However, it is a consequence of that theorem that there exist infinitely many arithmetical objects, and, traditionally, analytic statements should have no existential consequences. Boolos (1997) challenged the analyticity of HP on these lines, essentially equating it to an Axiom of Infinity. Neologicists have replied (Wright 1999) that HP, *by itself*, has no existential consequences: being a biconditional, it only equates the truth-conditions of its two sides, but is silent on whether those of the RHS are ever instantiated (something which depends on the background logic). It is, as it were, a purely nominal definition of $\#F$-terms. Secondly, neologicists contend, the order of explanation should be reversed. It is not that if HP has existential consequences, it is not analytic; rather, if HP satisfies criteria for being an acceptable definition, and it turns out to have existential consequences, then it is the traditional conception of analyticity as existentially neutral that should be revised – a move which, after all, is wholly consistent with Frege's revision of that notion.

Neologicism, if successful, provides a rationalist (or "intellectualist": Hale and Wright 2002) reply to Benacerraf (1973)'s challenge: it accounts for arithmetical knowledge and access to arithmetical objects *a priori*, via standard reasoning capacities, starting from linguistic knowledge. Crucially, especially for present purposes, it is based on the role of a unique fundamental definition. It is then worth surveying briefly some of the many concerns affecting implicit definitions in this program.

> *Sol.* Frege Arithmetic relies on second-order logic, and HP is formulated in a second-order language. Neologicists must address Quinean doubts on the legitimacy of second-order logic (Wright 2007). Other authors suggest adopting plural quantification (i.e. quantification over pluralities rather than sets) to escape the apparent set-theoretical commitments of SOL; these views could then be adapted to logicist accounts Linnebo 2022, §4).

> *Reconstructive Epistemology.* As a reconstruction of an *a priori* route to arithmetical knowledge (and not a cognitive or developmental description of how such knowledge is actually formed), neologicist epistemology is prey, among others (Nutting 2018), to the same objections leveled against **Rational Reconstruction** (see Section 4.2). One concern is whether it must be taken as an *hermeneutic* project, aiming to capture what subjects meant all along when using the concept of number, or a *revolutionary* one, advancing a theoretical replacement for that notion. Frege Arithmetic may in fact provide a theory which is just an isomorphic translation of arithmetic, rather than arithmetic itself (Heck 2000). This challenges the role of HP for **Conceptual Analysis**, since the neologicist concept of cardinal may fail to capture adequately, or uniquely, the informal notion it targets. The possibility of giving different abstractive reconstructions of the same informal notion can then either lead to pluralist interpretations of abstractions (see Section 7.4.1), or to acknowledge that far from delivering univocal analysis or reductions, abstractions are rather the outcome of a much more nuanced process of explication (see Section 6.2; Reck 2007).

SYNTACTIC PRIORITY THESIS. Neologicists appeal to a version of the Context Principle, the *Syntactic Priority Thesis*: if an expression qualifies, by purely syntactic criteria, as a singular term, and there are reasons to hold true a class of purportedly referential (nonopaque) contexts in which it occurs, nothing else is required for the expression to have genuine objectual reference. Provided #F-expressions behave syntactically as singular terms, they can be said to refer to objects, and the latter can then be proved to exist, based only on the truth of instances of the RHS of HP. Finding purely syntactic criteria for singular-termhood, however, isn't straightforward (Hale and Wright 2001a, Essays 1,2; Schwartzkopff 2016). Also, the metaontology underling Neologicism (Hale and Wright 2009b), according to which the existence of objects can be guaranteed by purely linguistic considerations, can be disputed.

IMPREDICATIVITY. Frege Arithmetic adopts unrestricted, namely impredicative SOL. Also, HP is impredicative insofar as cardinals are objects populating the first-order domain of its RHS. According to Dummett (1991), ch. 17, in order to grasp a first-order quantified statement we should be able to have a complete grasp of its domain of quantification. But this can't be done for the RHS of HP, for its domain contains objects that can be acknowledged only once HP has been advanced. Predicative restrictions to HP can be explored (Linnebo 2016). Still, neologicists believe that impredicativity poses no real epistemological challenge (Wright 1998), since a subject going though the steps of Frege's Theorem can start with a concept ($\ulcorner x = 0 \urcorner$) which requires no grasp of objects, and then only requires, at each step, to have a grasp of the cardinal numbers introduced in the previous steps, but never of the entire first-order domain – making impredicativity epistemically "harmless."

BAD COMPANY. All abstractive definitions should be *a priori*, consistent, and jointly consistent. But BLV isn't. The *Bad Company Objection* (Dummett 1998) states that HP cannot be a good definition since it shares the same logical form with bad abstractions. Neologicists have turned this into a request for criteria for acceptable abstractions. One is CONSISTENCY, which rules out BLV. But this is not enough. The so-called Nuisance Principle, NP (Wright 1997, 289–290) states that two concepts have the same Nuisance *iff* they differ finitely – *iff* the concepts $\ulcorner F \wedge \neg G \urcorner$ and $\ulcorner \neg F \wedge G \urcorner$ are both finite. NP is a consistent abstraction, but has only finite models and is then jointly unsatisfiable with HP, which is only satisfiable in infinite domains. Both HP and NP fail NONCREATIVITY/CONSERVATIVENESS. On that constraint, $\{T + HP\}$

should not entail sentences expressible *in the language* of T which are not already entailed by (or derivable from) T alone, for any T. But this HP does, since it entails ϕ = 'There exist infinitely many objects', which is expressible in T but may not be a consequence of T alone if T has a finite domain. However, while NP entails that the entire domain (including the domain of T) must be finite, HP only entails ϕ with respect to its own ontology of cardinals, remaining silent on the domain of T alone. Hence, it satisfies the following constraint:

> SEMANTIC FIELD-CONSERVATIVENESS: an abstraction principle ABS is *semantically Field-conservative* if for any theory T to which ABS can be consistently added, and for any sentence $\phi^{\neg\Sigma}$ of the language of T whose quantifiers have been restricted to the ontology of the original theory, $\{T + ABS\} \models \phi^{\neg\Sigma}$ only if $T \models \phi^{\neg\Sigma}$.

Still, it is possible to find pairs of abstractions which are consistent, semantically Field-conservative, and consistent with HP, but are pairwise inconsistent since they are satisfied only if the domain of objects they introduce are of different infinite cardinalities. Further constraints are needed, but other bad companions are to be expected, and the list of constraints (such as irenicity, stability, etc.) must then be expanded (see e.g. Linnebo 2009; Cook and Linnebo 2018; on conservativeness, see also Cook 2012; Mackereth (in press)).

GOOD COMPANY. As Mancosu (2016) shows, different equivalence relations on the RHS of HP yield different *good* cardinal abstractions. Frege (as Dedekind and Cantor) measures cardinalities in terms of equinumerosity. As a consequence: (*i*) a set x which is a proper part of a set y can have the same infinite cardinality; (*ii*) nonequinumerous infinite sets have different infinite cardinalities (naturals and reals, etc.). We can replace equinumerosity with Peano's cardinality function *num*(a) and obtain an alternative Peano's Principle (PP): *num*(a) assigns cardinals to finite concepts in the same way as HP, but every infinite concept is assigned the same infinite cardinal (∞), against (*ii*). Peano's Principle also ignores part–whole relations (infinite sets which are proper parts of infinite sets are always assigned the same cardinal). Yet another principle, Boolos's Principle (BP), assigns finite cardinals to finite concepts like HP, but then assigns a cardinal a to each concept F which is both infinite and coinfinite (i.e. such that $\neg F$ is also infinite); and a different cardinal b to each concept G which is infinite but cofinite (i.e. such that $\neg G$ is finite). PP and BP satisfy criteria against bad abstractions and derive the PA^2 Axiom. Yet, we ultimately end up with different definitions of cardinal number, diverging on assignments to infinite concepts. But how to tell then which one is the correct one? This exacerbates for neologicists the problems afflicting **Rational Reconstruction** (see also Section 7.4.1).

Extending Abstraction: Real Numbers

Real numbers can be easily defined axiomatically. Their structure is a complete ordered field. A field is an algebraic structure constituted by a set with an operation of addition + and multiplication × defined on them, and an additive and multiplicative inverse. Apart from field axioms and order axioms, a Completeness Axiom establishes in which sense reals differ from discrete sets (e.g. $\mathbb{N}$) and dense sets (e.g. $\mathbb{Q}$): every nonempty set of reals which is bounded above has a least upper bound. This axiomatic definition characterizes directly the real-number structure. Alternatively, however, it is possible to define the reals as particular objects whose properties will then ensure that they satisfy completeness and total order and that field operations can be defined on them. This could be done by abstraction too.

In Frege (1893–1903), II, Frege gestures at how to extend his program to the reals (Dummett 1991, ch. 22; Simons 1987). A full treatment was left to a third volume, which was never published in light of Russell's Paradox. Today we know of different ways to define reals by abstraction. Shapiro (2000) provides one. He first defines naturals via HP, and then, by repeated abstractions, integers from naturals, quotients of integers from integers, and the reals by means of an abstraction which essentially introduces Dedekind's cuts. This procedure, like the genetic method discussed earlier, ultimately bases the definition of the reals on the naturals. Hale (2000) advances a Cut Abstraction closer to Frege's thoughts. We know Hilbert rejected the genetic method and favored axiomatic presentations. Frege too rejects it, but for different reasons. He believes that defining reals based on numerical constructions betrays their essential purpose of expressing ratios of quantities, a feature – crucially related to their applications in measurement (Section 7.6.2) – that should be captured in their definition. Hale's Cut Abstraction defines reals as ratios of quantities defined on suitably specified quantitative domains. The existence of at least one such domain is granted by the domain of ratios (introduced by abstraction) among naturals (introduced by HP). Other options are available, and one open question is whether once properly reconstructed, a Fregean definition of the reals is really able to sustain the extension of Frege's logicist views to analysis (Boccuni and Panza 2022). Even harder issues arise as regards the possibility of extending the neologicist program to set-theory (Cook 2021). The possibility of extending neologicist abstractionism beyond arithmetic is essential in comparing the prospects of this program to those of axiomatic approaches in recovering higher branches of mathematics (and contributes, more generally, to a wider debate on a general framework for abstractive definitions and their merits; Ebert 2016; Panza 2016).

6 Elucidations and Explications

Definitions are properly effected within the language of a given theory. But, as we have seen, in most interesting cases, and especially when primitive notions are involved, definitions are the terminating end of a process of analysis which starts from informal notions coming from previous mathematical practice. There are several ways of explaining how this process occurs. Although we have already hinted at related issues, it is worth pausing a little on some available options.

6.1 Elucidations

Theories must start somewhere. Explicit definitions (as well as implicit definitions of nonprimitive terms) must be given within a theory by relying on its basic vocabulary. Unless we devise a way (axiomatic, abstractive, or other) to provide something akin to a definition of such basic vocabulary, primitives cannot be defined, and still their meaning must be conveyed somehow if we are to understand the subject matter of the theory at all. *Elucidations* provide a way of accomplishing this. These generally point to processes of illustrating the meaning of a concept by means of a previously known language, when no possibility of further reduction or definition is available. In a sense, as we have seen (see Section 2.3), the primitives of Euclid's original presentation in the *Elements* are elucidated rather than defined. They are introduced by statements of whose meaning an understanding must be granted, lacking any further reduction to more basic concepts, in virtue of how those notions are used before theory construction.

Elucidations can thus delineate the meaning of primitive notions before a formal treatment is provided. In Frege (1893–1903), I, §§34–35, Frege suggests that elucidations (*Erläuterung*) are ways of initially conveying the intended meaning of the primitive notions of his formal language, without expecting completeness or full exactness (see also Frege 1979a, 207). Concerning geometrical primitives too he states that "we must admit logically primitive elements that are indefinable" (Frege 1971, p. 59) and that can only be elucidated. Elucidation, in this sense, is no proper part of a scientific theory: "it […] serves the purpose of mutual understanding among investigators, as well as the communication of the science to others," and its role is mainly an heuristic or "propaedeutic" one with the aim of "mutual cooperation" (Frege 1971, p. 59). Precise definitions are only available from within. It is again on this ground that Frege criticizes (p. 60) Hilbert's attempted axiomatic definitions of the geometrical primitives, because they are not presented as mere elucidations, but they are "cornerstones of the science" and can be used "as premises of inferences" within the theory. Also, a pivotal role of elucidation is to let investigators converge on a specific and univocal meaning of the relevant terminology, and we have already seen that one of the features (which is, in Frege's eyes, a defect) of Hilbert's axioms is that different interpretations of the primitives are equally admissible.

6.2 Explications

Although the terminology is sometimes overlapping and the German '*Erläuterung*' is translated differently, we should distinguish elucidations and *explications*. Explications are ways of giving a formal rendition of informal notions even though their analysis may fail to be faithful to every aspect of common usage. They are a way in which **Conceptual Analysis** may proceed (Carnap 1947, I.2) and they result in analyses that are not merely antecedent or propaedeutic to the theory, but form a proper part of it. As Carnap (1950b), p. 3, puts it in introducing his logical analysis of the notion of probability:

> The task of explication consists in transforming a given more or less inexact concept into an exact one or, rather, in replacing the first by the second. [...] The explicandum may belong to everyday language or to a previous stage in the development of scientific language. The explicatum must be given by explicit rules for its use, for example, by a definition which incorporates it into a well-constructed system of scientific either logicomathematical or empirical concepts.

Carnap extensively elaborated this method (Novaes and Reck 2017; Reck 2024), and advanced explicit requirements for explications (Carnap 1950b, §I.3): (a) *similarity*: the *explicatum* should be as similar as possible to the *explicandum* (complete correspondence is obviously not required or even attainable); (b) *exactness*: it should be given precise rules of use; (c) *fruitfulness*: it must help deliver as many theorems (or scientific laws) as possible; (d) *simplicity*: it must be expressible in the simplest form compared to other possible choices of *explicata*. Indeed, the process is by its very nature comparative (and hence closely related, in both merits and limits, to **Rational Reconstruction**): "[...] if a solution for a problem of explication is proposed, we cannot decide in an exact way whether it is right or wrong. [...] The question should rather be whether the proposed solution is satisfactory, whether it is more satisfactory than another one, and the like."

As examples concerning logico-mathematical concepts, Carnap mentions Tarski's (1933/1956) formal analysis of truth, Menger's (1943) treatment of the notion of dimension, and, most notably for us, a combination of Peano's axiomatic and the Frege–Russell analysis of cardinal number (the former as a way of providing an exact axiomatic rendition, the latter as a way of settling on one particular interpretation of those axioms among the many available equivalent progressions).

Explications, however satisfactory they eventually are, must then sacrifice something of the original usage of the informal notion. Quine (1960), §53, takes the notion of ordered pair as a "philosophical paradigm" of this kind of analysis. Different explications of the pretheoretical notion of ordered pair, as for example $\{\{x\}, \{x, \emptyset\}\}$ or $\{\{x\}, \{x, y\}\}$, provide formal renditions with the required set-theoretical properties, but do so at the expense of adequacy to pretheoretical informal use. Similar remarks apply to Russell's theory of definite descriptions, or to Frege's analysis of cardinals as equivalence classes. In all these cases:

> We do not claim synonymy. We do not claim to make clear and explicit what the users of the unclear expression had unconsciously in mind all along. We do not expose hidden meanings, as the words 'analysis' and 'explication' would suggest; we supply lacks. We fix on the particular functions of the unclear expression that make it worth troubling about, and then devise a substitute, clear and couched in terms to our liking, that fills those functions. (*ibid.* 257–258)

In this sense, "explication is elimination" (259), in that the original notion is discarded, and the differences with the newly introduced one are ignored.

The method of explication provides further nuances as to how **Conceptual Analysis** can be understood. On one conception, which may be suited to traditional foundational project, analysis targets an informal notion that, however underspecified or vague in its ordinary usage, is meant to be unique and uniquely characterizable. One way of contrasting this picture is to claim that the target notion is indeed unique bust still schematic, and allows for a plurality of specifications (see Section 7.4.1). Explication seems to renounce the very idea that the outcome of analysis (either in a monist or pluralist sense) can be uniquely determined. Pragmatic factors, including issues concerning the most appropriate systematization of our scientific theories, enter the comparative evaluation of alternative analyses. This picture can adequately account for the actual development of alternative formal theories of the same informal notion, a most notable example being the various axiomatization of the notion of set (see Section 7.3.2). Axiom systems (as well as abstractive definitions) can provide the required exactness to the *explicatum*. But different (theoretical, mathematical, and philosophical) considerations may lend support to one or another system.

To a certain extent, this method is faithful to how we clarify both ordinary and scientific concepts, and how different clarifications are offered over time. The nature and details of this process of *conceptual engineering* may vary. It affects both ordinary and socially relevant notions (*viz.* gender, or family), and scientific ones (*viz.* mass, light, or planet). Arguably (Tanswell 2018), we also conceptually engineer logical and mathematical notions: think of how different conceptions of logical validity are motivated by nonclassical logics, or how conceptions of number or set have been differently explained or just expanded through time. Explication and conceptual engineering, then, bear resemblances worth exploring (Brun 2016, 2020) – and in the end are both related to the method of reflective equilibrium mentioned in Section 2.1.

A pressing issue in either case – something that was already foreshadowed earlier in discussing the Frege–Hilbert debate – is whether change in concepts entails a replacement view, that is, a change of subject, so that for example users of the two concepts would not actually disagree, but simply talk past each other about different things (Quine 1970, §6); or whether some continuity of topic can be granted, so as to take the concept resulting from the analysis as an ameliorative version of the original one.

7 Knowledge by Definition

The preceding general taxonomy of definitions and their philosophical roles leaves many open questions as to whether, and how, definitions of different sorts can provide mathematical knowledge. This section briefly outlines some of these major issues to provide a map of the recent and current debate.

7.1 Definitions and Empirical Evidence

7.1.1 Naive Empiricism

If, as empiricists would have it, all knowledge is to be grounded on empirical evidence, knowledge provided by definitions should be too. Views on these lines were advanced by Mill (1843) (and von Helmholtz). Mill conceives of arithmetical laws (and geometrical postulates) as inductive generalizations on observed experience, rather than *a priori* self-evident propositions. As such, they are contingent and can be falsified by experience, although we may be psychologically unable to conceive it. As a consequence, he takes the notion of sum to generalize instances of unions of physical unit quantities, and believes that

1+1 does not necessarily result in 2, for it is always possible that, by physical processes, in adjoining one and one unit of a given substance, less than two units result. In this setting, definitions assert observed matters of fact. For instance, the definition of '3 $=_{df}$ 2 + 1' would assert the possibility of rearranging a collection of objects, say three pebbles, so that they can either impress our senses as ⊚⊚ ⊚ or as ⊚⊚⊚. Numerical terms denote particular physical collection, and connote their physical properties. To these views, Frege harshly objected that empiricists confuse "the applications that can be made of an arithmetical proposition […] with the pure mathematical proposition itself" (Frege 1884, §9; see also Frege 1893–1903, Vol. II, §137) – on the other hand, Mill's view that apparently obvious and indubitable truths can possibly be false was much more aligned with subsequent developments in non-Euclidean geometries.

Views like Mill's are generally considered untenable (but see Kitcher 1983). A more nuanced form of empiricism only claims that mathematical statements should be justified by empirical means (not that their content is empirical), on the basis of the role they play for the expressive or explanatory power of scientific theories – as the so-called *Indispensability Argument* suggests (Panza and Sereni 2013, chs. 6–7). Here, however, definitions play little epistemological role by themselves, and are rather confined to how the background mathematical theories are constructed.

7.1.2 Cognitive Adequacy

More recently, however, studies in the cognitive sciences and neurosciences greatly incremented our appreciation of the cognitive and neurological underpinning of numerical abilities in humans and animals (Carey 2009; Dehaene 2011; Samuels and Snyder 2024). Neuroscientific studies on mathematical cognition could be used to dismiss philosophical concerns. As Quine (1969) rejected traditional epistemology in favor of psychology, one may dismiss philosophical (epistemological, normative) concerns on mathematics in favor of scientific (data-driven, descriptive) inquiry on numerical capacities. An alternative, more conciliatory attitude consists in asking foundational theories to take neuroscientific evidence into due account. As regards definitions, one may even add a further methodological constraint:

COGNITIVE ADEQUACY. Definitions of numerical (or mathematical) concepts in philosophical reconstructions of mathematical theories should be cognitively adequate, that is, should cohere with cognitive and neuroscientific evidence on the possession, formation, and development of such concepts.

Whether this would be a desirable constraint is controversial. Even without being dismissive toward either neuroscientific or philosophical inquiries, one can maintain a healthy division of labor. Philosophy and neuroscience are legitimate but distinct enterprises, differing in methodology and aims: the search for justification and the description of concept formation should not be confused.

If one endorses the criterion, however, one may wonder which, among alternative definitions available of, for example, natural number, is most cognitively adequate. A significant challenge on this score is that neuroscientific evidence on numerical cognition still doesn't settle the nature of our basic number concepts. Some evidence suggests that our numerical abilities are grounded on elementary appreciation of order among small collections of objects: a structuralist conception, or anyway a conception of natural numbers as essentially ordinals, would thus seem preferable. Other evidence acknowledges an essential function to the ability of immediately grasping the numerosity of small collections of objects (subitizing), which would better cohere with a definition of natural numbers along (neo-)Fregean lines, or anyway as essentially cardinals (Decock 2022).

Be that as it may, one particular difficulty consists in merging empirical research on very elementary numerical skills with the epistemological concerns regarding complex and mature mathematical theories. A yet harder question, which has been tackled recently (Pantsar 2024), is whether it is possible to provide an account of mathematical (or at least arithmetical) knowledge that is attentive to traditional epistemological concerns while being empirically informed on both the neurocognitive basis of numerical abilities and the impact of training and society (enculturation) in the development of numerical concepts.

7.2 Knowledge without Objects

7.2.1 Conventions

Some believe that language is not in the business of capturing objective, independent constituents of reality. Rather, language use is constitutive of our concepts, and which concepts we adopt to categorize reality is partly arbitrary. Definitions then act as *conventions*, which we are free to stipulate at our convenience for various purposes, and still secure *a priori* knowledge. A major inspiration comes again from Carnap, who neatly distinguished between questions internal and external to a given linguistic framework (Carnap 1950a). External questions ("Are there really numbers?") are only to be settled by pragmatic reasons, for example by the convenience of adopting a linguistic framework in which specific kind of entities are contemplated. Internal questions ("Are there prime numbers?") are instead to be addressed by considering what follows internally to a linguistic framework from its basic principles, including meaning postulates establishing the meaning of basic expressions.

The choice of postulates is entirely conventional (although conventions can be compared for their usefulness and fruitfulness as the outcome of alternative explications).

As anticipated, the possibility of truth by convention is essential to logical empiricism (see p. 55). Together with analyticity and synonymy (Section 5.2.2), the idea has been challenged by Quine (Quine 1936, 1954), although it has recently been defended again (Warren 2020).

7.2.2 Fictions, Creations, and Constructions

A radical distaste with an ontology of abstract objects underlies nominalist views. We have seen how eliminative definitions can sustain nominalism. They can also support fictionalist varieties of nominalism. Fictionalists' major contention is that any apparent reference to mathematical objects should be seen as a *façon de parler*, an helpful instrument with no proper referential role. Definitions can thus be seen as *fiction-introducing principles*. For instance, once we have numerical quantifiers as defined in Section 3.2, we can introduce numerals, grammatically behaving as singular terms, as fictional tools for simplifying lengthy deductions ('$\#F = 2$' will replace the long expansion of '$\exists_2 x F(x)$'). Field (1980/2016) argues along these lines. Yablo (2005) advances a related view, in which definitions could be conceived as bridge principles, establishing how to translate certain statements about the real world into figurative (or metaphorical) ones within a particular make-believe game, to "make as if" there were numbers.

Notice that fictional realists maintain that fictional terms genuinely denote (mind- and language-dependent) fictional objects. Within this framework, mathematical fictions would be about objects after all – and fictionalism would come close to forms of creation. Similar remarks apply to views somehow intermediate between creationism and conventionalism. Cole (2015), for instance, suggests that mathematical objects should be conceived as social constructions (and, arguably, definitions would be akin to social conventions). The view inherits all concerns affecting social ontologies generally, but has the merit of connecting mathematical ontology to a conception of mathematics as a human, historically and socially determined, practice (Cantù and Testa 2023).

7.3 Knowledge through Axioms

7.3.1 Consistency and Existence

If we believe that axiom systems are not mere conventions, but rather theoretical descriptions of certain mathematical domains, then how can we know they are true, and how can we have knowledge of the objects they are allegedly about? Recall a fundamental distinction between the traditional Euclidean model and Hilbert's conception. On the former, a prior, extratheoretical access to an external geometrical reality provides content to primitives and truth to postulates. That a straight line lies between any two points, and that points are objects of a given kind, is not something we know as a consequence of the theory, but something that can be ostensively presented to test its adequacy. The possibility of such extratheoretical evidence is forsaken in Hilbert's conception. Knowledge of both truths and objects can only be afforded by the theory. As Hilbert writes to Frege: "If the arbitrarily given axioms do not contradict one another with all their consequences, then they are true, and the things defined by the axioms exist. This for me is the criterion of truth and existence" (Frege 1980, 29.12.1899, 39, 42).

Similar views are today displayed in *plenitudinous* varieties of platonism (Balaguer 1998). Closer to a Euclidean spirit, Frege retorts that "from the truth of the axioms it follows that they do not contradict one another." Axioms are guaranteed to be true insofar as they are able to capture objects and properties that can be independently exhibited. We don't first have a theory and then look for a(ny) model of it (as in Hilbert's metamathematical proofs); we display a(n intended) model and then theorize about it. This holds for arithmetic too, where definitions provide the objects from whose properties arithmetical truths will follow. This almost ostensive exhibition of the intended model makes any additional consistency proof supererogatory (Hallett 2019, 301). For Frege, Hilbert's conception suffers the same limits of ontological proofs for the existence of God: from the joint consistency of claims that x is intelligent, omnipresent, and omnipotent, it does not follow that such an x actually exists (Frege 1980, 6.1.1900, 47). In axiomatics, existence must rather be presupposed, and cannot be proved. In Aristotle's terms, primitives need both a nominal definition and a presupposition that they exist.

An analogous dialectics is present in more recent debates. If we see axioms as defining *ante rem* structures, and structures are sets of (possibly uninstantiated) relations, it's less controversial to take the existence of structures as entailed by the consistency of axioms (or by their coherence, a narrower notion adopted by Shapiro 1997). However, neologicists (Hale and Wright 2002, 112–113) follow Frege in seeing here an unwarranted passage from "conveying a concept" to "in addition, induc[ing] awareness of an articulate, archetypal object": conceptual understanding *alone* cannot provide knowledge of objects.

7.3.2 Choosing the Axioms

If primitive concepts are entirely determined by axiomatic definitions, then a change in axioms is a change in concepts. As Hilbert already claimed: "only the structure of axioms yields a complete definition. Every axiom contributes something to the definition, and hence every new axiom changes the concept" (Frege 1980, 29.12.1899, 40, 42).

Modifying axioms isn't problematic if we take axioms as arbitrary stipulations, but it is if axioms are meant to provide univocal analyses of prior notions. An illuminating example is offered by axiomatic set-theory. Axiomatizations like ZF and NBG advanced in the early twentieth century could be seen as formal reconstructions of an informal notion of set. To avoid paradoxes, they include formal conditions going well beyond what seems implicit in pretheoretical usage (respectively, the idea that sets form a *cumulative hierarchy*, and that sets and proper classes should be distinguished by a principle of *limitation of size*; Potter 2004; Ferreirós 2007; Barton 2024). A suitable choice of axioms is crucial whenever the analysis proves somehow incomplete, as revealed for example by the recognition that ZFC does not decide the Continuum Hypothesis (CH) (namely that the cardinality of the set of real numbers is $\aleph_1$, hence that there is no intermediate infinite cardinality between the cardinality of countable sets like natural numbers, $\aleph_0$, and that of the reals, $2^{\aleph_0}$). It follows that both ZFC+CH and ZFC+¬CH are consistent. Additional axioms can decide CH one way or another, but how do we choose? One can either rely on *extrinsic*, possibly inductive, evidence for purely mathematical or architectural reasons; or on *intrinsic* evidence directly relating to our pretheoretical notion of set, believing (like Gödel 1964) that our initial axioms failed to capture it completely (Maddy 2011; Linnebo 2017, ch. 12). The debate on alternative axiomatizations of set-theory in light of higher set-theoretical principles has flourished recently (Antos, Friedman, Honzik, and Ternullo 2018). If one renounces the idea that only one analysis is to be correct, axiom revision comes close to processes of explication or conceptual engineering (see Section 6.2).

7.4 Knowledge of Objects

7.4.1 Multiverses, Good Company, and Pluralism

Maybe, by changing the axioms we are neither replacing nor improving old concepts, but rather providing rival and still equally acceptable specifications of the same concept. We renounce UNIVOCALITY not because different analyses are available for comparison, but because the very notion we are analyzing proves somehow schematic and open to a plurality of specifications.

This form of *pluralism* has been extensively explored in logic. Among different varieties of logical pluralism (see Shapiro 2014), Beall and Restall (2005) argue that rival logics (provided suitable criteria are met) are equally legitimate outcomes of alternative specifications of a general schematic definition of logical validity. This view could be transferred to mathematical concepts too. Different axiomatizations of set-theory could provide different and still legitimate specifications of a common, underdetermined if not schematic, concept of set. A controversial consequence ensues anytime the target concept is a concept of objects and hence a form of *ontological pluralism* is engendered. If any axiomatization characterizes a set-theoretical universe, the adequate ontological view could be that of a *Multiverse* of sets, a plurality of coexisting, distinct although possibly overlapping set-theoretical domains (Hamkins 2012).

Mathematical pluralism can be defended in various ways (Priest 2024), and can also be sustained via abstractive definitions. Recall the *Good Company Problem* (Section 5.2.2), according to which a plurality of equally acceptable cardinal abstractions is available. A possible way to interpret this result (Sereni, Sforza Fogliani, and Zanetti 2023) is to see the informal notion of cardinal as underdetermined, and *all* good companions as legitimate ways of specifying it. In analogy to logical pluralism, these could be seen as based on different specifications of a schematic relation of 'being the same size as' that the RHS of HP should capture. Different cardinal abstractions would describe a plurality of co-existing, partially overlapping (on the finite) and yet distinct numerical domains.

7.4.2 Real Definitions and Metaphysical Foundationalism

Objectual knowledge through definitions is, at least, indirect: it is propositional knowledge *that* some objects exist and are so-and-so. However, one may follow Aristotle and also ask for knowledge *why* mathematical objects are, on what their existence and essence are grounded, by taking mathematical definitions as real definitions. Some proposals with a clear metaphysical focus take mathematical definitions as either individuating the essence of numbers as necessary beings (Hale 2013, ch. 6 and 7), or encoding all and only those properties that establish numbers as particular abstract objects (Zalta 1999; Nodelman and Zalta 2024). Some recent proposals take the relation between *definiendum* and *definiens* as an (asymmetrical) relation of metaphysical reduction (Rosen 2010). While in an explicit definition the *definiens* establishes just what the *definiendum* is (it metaphysically identifies the latter with the former), in implicit definitions metaphysical reduction relates not just the target entities, but facts involving them. This brings up notions pervasive in analytical metaphysics like *grounding* and *ontological dependence*.

Grounding (Fine 2012; McKenzie 2022) is a relation of metaphysical, non-causal explanation between facts (or, on some views, between statements). It is expressed by 'in virtue of' or 'because' idioms, and underlies claims such as

"The fact that the singleton set of Socrates exists obtains in virtue of the fact that Socrates exists." It is formalizable as '$A \prec B$' ('A *grounds* B'). As all explanatory relations, it is asymmetric ($A \prec B \rightarrow B \nprec A$). Strict ground is irreflexive ($A \nprec A$), in contrast to weak ground (where $A \preccurlyeq A$). On most conceptions, it is factive ($A \prec B \rightarrow A, B$) and transitive ($A \prec B \wedge B \prec C \rightarrow A \prec C$). A fact A can ground a fact B alone (full ground) or together with other facts (partial ground). Ontological dependence can loosely be seen as expressing the relation 'x exists in virtue of the existence of y.'

Traditional foundational programs developed in a milieu permeated by a distaste for traditional metaphysics and a priority assigned to semantic analysis. As Dummett (1993b), 6, epitomizes this: "Language may be a distorting mirror: but it is the only mirror that we have." Still, several authors acknowledge that even when Frege characterizes foundations as "afford[ing] us insight into the dependence of truths upon one another" (§2.1), his "relation of grounding or dependence is as much metaphysical as it is epistemic" (Shapiro 2012, 221–222). As Wright (2020), 283, has it, Frege's claims are "suggestive of a kind of metaphysical architecturalism" and "Frege never shakes off a broadly Euclidean view" to the extent that "the axioms should codify the *ground* of the truth of the theorems, the body of fact on which those truths rest." After all, considerations on the essence of arithmetical objects surface in both Fregean, neologicist, and structuralist perspectives. Can metaphysical insights be afforded by either axiomatic or abstractive definitions?

In a structuralist framework, one can wonder which dependence holds between objects and structures (Linnebo 2008). On the one hand, it seems that relations cannot hold without *relata*: the existence of structures depends ontologically on the existence of their instances. However, in an *ante rem* setting, *relata* are just places (or "offices") in structures (although conceivable as objects; Shapiro 1997, ch. 3), and these are determined by the structure: objects, that is, places, should rather be seen as depending ontologically on structures. Also, since positions in a structure are determined by their relation with one another, positions may depend for their identity on all others.

What about abstractions? One can take the consequences of Frege's theorem, arithmetical axioms in particular, as facts grounded on logical laws and HP. However, while logical derivation may provide an epistemic dependence between premises and conclusions, one could take all arithmetical facts to be metaphysically on a par and equally fundamental. A different issue is whether the LHS of HP can be seen as metaphysically grounded on its RHS, that is, if identity of cardinals holds in virtue of equinumerosity of concepts. However, on some conceptions of content-recarving, abstractions provide different conceptualization of *the same* fact, differently described on the two sides. As Hale and Wright (2009b), 187, claim, "the existential commitments of the statements which the abstraction pairs together *are* indeed the same" (which, in passing, seems to require the RHS to effect *hidden reference* to cardinals even though no surface denotational device is available). Equinumerosity of concepts *is* (although it is said differently) identity of cardinals (see also Section 7.5.1). It is then hard to see how the two sides could be assessed asymmetrically. Finally, one may ask whether the objects whose existence HP helps to prove are ontologically dependent on one another. Various proposals have been advanced (Rosen 2010; Schwartzkopff 2011; Donaldson 2017). However, they outline a hierarchy of dependence relations which appears in tension with a platonist setting, and is better located within a renaissance of neo-Aristotelian views in which mathematical existence is grounded on more fundamental, possibly nonmathematical, facts (Panza and Sereni in press). Other recent proposals (deRosset and Linnebo 2023) apply grounding to other varieties of abstractionism (in particular, to Linnebo's conception of thin objects; see Section 7.5.1).

7.5 Knowledge, for Little or Nothing

7.5.1 Thinness

The concern with the existential consequences of definitions, abstractive ones in particular, is partly due to a traditional conception of platonism: abstract objects seem to impose too heavy a metaphysical burden to be warranted only through our conceptual and linguistic procedures.

One route is to claim that while the notion of reference in most areas of language (e.g. empirical discourse) is in some sense robust, abstractions only secure a *thin reference*. Dummett (1973), 223–224, distinguishes between realistic reference and semantic role, the first being modeled on the name/bearer relation, the latter being the contribution that an expression makes to the truth conditions of a statement. Since in an abstraction what is at stake is essentially the equivalence between the RHS and LHS, "the notion of the reference of the [terms in the LHS] [...] plays no role in our conception of what determines the thought as true or false" and is thus "semantically idle" (Dummett 1991, p. 193). This view is, however, hostage to an explanation of how reference happens to behave differently in different areas of language.

Without revising the notion of reference, it is then possible to defend the Fregean or neologicist approach to ontology as grounded on semantical considerations. The basic insight (coherently with the linguistic turn inaugurated by Frege) is that the picture according to which we need a prior access to, or conception of a given sort of objects, in order to be able to understand thoughts about them, is misguided. In relevant cases the order of explanation is reversed, and it is propositional knowledge (e.g. that involved in the stipulation of an abstraction principle) that gives us all we need (and can have) to obtain knowledge of certain objects and of statements concerning them (Hale and Wright 2009b).

Various strategies, despite differences, agree in claiming that the kind of ontological knowledge provided by abstractions is, in some sense, unproblematic or anyway *easy* to obtain (Thomasson 2014). One possibility is to claim that the very nature of the objects involved is such that it makes no substantial metaphysical demands on the world. For instance, Rayo (2013) considers *just-is* statements, that is, statements stating that, for suitable statements p and q, for p to be the case *just is* for q to be the case, so that the ontological import of the two statements is the same (e.g. for the number of dinosaurs to be 0 just is for there to be no dinosaurs). Reading abstractions as just-is statements could provide a form of *trivialism* that defuses their metaphysical robustness by claiming for a strong symmetry between its two sides, namely that for two numbers to be identical *just is* for two concepts to be equinumerous. Linnebo (2018b) has developed a distinct form of *minimalism* based on the notion of *thin objects*, namely objects whose existence does not make a substantial demand on the world. Linnebo reads abstractions as sufficiency statements, that is, statements establishing that asserting the RHS gives sufficient grounds for asserting the LHS. This affords an asymmetric reading of abstraction, and the ontology of the LHS is a genuine addition with respect to that of the RHS. The RHS gives identity conditions which assign genuine objectual referents to the terms in the LHS, whose demands on the world (their conditions of existence), however, do not exceed those expressed in the RHS, ensuing in another variety of *lightweight* platonism (Sereni and Zanetti 2023). These sufficiency claims extend both the original language and the original ontology, in a process of *dynamic abstraction* with several ramifications (Carrara and Zanetti 2023).

7.5.2 Entitlement

The problem of obtaining noninferential justification of basic principles has special significance when applied to basic logical laws too. The *centrality* of most basic deductive (and inductive, as Hume showed) logical principles entails that any attempt at their justification is bound to circularly appeal to the very same principles. Cognate remarks apply to a larger family of statements which are central to many cognitive enterprises but whose inferential justification is unattainable. For instance, skeptical challenges suggest that any inferential justification for the claim that there exists an external world will be a case of *transmission failure*: it will fail to transmit justification from its premises to the target conclusion by surreptitiously relying on the very fact that it wants to establish, namely that an external world exists.

All these cases can be seen as instances of what Wittgenstein (1969) calls "hinge" or "cornerstone" propositions: propositions so deeply entrenched in our conceptual schemes as to be constitutive of them, and as such impossible to be justified from within. Relying on them requires alternative forms of epistemic warrant. Wright has appealed to the notion of *epistemic entitlement* (Wright 2004a, 2004b). Entitlement is in place whenever a presupposition P underlying a specific cognitive project is such that "(*i*) There is no extant reason to believe not-P and (*ii*) Someone pursuing the relevant project who accepted that there is nevertheless an onus to justify P would thereby implicitly commit themselves to an infinite regress of justificatory projects, each concerned to vindicate presuppositions of its predecessor" (Wright 2016, 166).

When we find ourselves in this situation, even in the absence of an inferential justification for P, we should take ourselves to be nonetheless entitled, or warranted (by default, or for nothing), in believing P. This holds for HP too. The epistemology of HP (and possibly other basic abstractions) would be analogous to that of basic logical rules, analogously to Gentzen-style definitions of logical constants. These can be provided by pairs of introduction and elimination rules. The meaning of e.g. '∧' is entirely exhausted by:

$$\frac{A \quad B}{A \wedge B} \wedge\text{-I} \qquad \frac{A \wedge B}{A} \wedge\text{-E}^1 \quad \frac{A \wedge B}{B} \wedge\text{-E}^2.$$

HP may be understood analogously (Hale and Wright 2000). Being in fact the conjunction of two conditionals, it works as a shorthand for an introduction rule (right-to-left) and an elimination rule (left-to-right) of the numerical operator '#_' (see also Tennant 1987). If Gentzen-style rules provide warrant, so does HP.

Entitlement is a much debated notion (Pedersen 2016; Graham and Pedersen 2020). If tenable, it would importantly qualify the kind of knowledge that can be afforded by HP. It preserves the possibility of attaining some form of *a priori* knowledge by implicit definition, but locates our belief in HP (leaching, possibly, to that in its arithmetical consequences) far from the "realm of apodictic certainty" (*Ibid.*, 183) usually associated with the *a priori*, seeing it rather as groundless and defeasibly warranted.

7.6 Axioms or Abstractions?

7.6.1 Arrogance

The epistemological significance of implicit definitions mostly relies on their *traditional connection* (Hale and Wright 2000) with *a priori* knowledge. To an extent, both abstractive and axiomatic definitons are well equipped to provide *a priori* an understanding of the concepts they define. For abstractive definitions, this is provided by the biconditional statement itself. For axiomatic definitions (for mathematical terms as well as theoretical terms in the empirical sciences) this can be done by taking the Ramsey-sentence (see Ramsey 1931) of a theory T, that is, roughly, the existential generalization over a conjunction of the axioms where primitives are replaced by suitable variables (so if the matrix of T's axioms is $\#(x)$, the Ramsey-sentence of T would be '$\exists x(\#x)$'); and then stipulating the corresponding so-called Carnap conditional: if there exists a t that satisfies '$\#x$', then '$\#t$' – something whose truth does not depend on the existence of any such t, and hence can be freely stipulated.

However, in both cases the truth of the definition itself (certain axioms or a certain abstraction) have, directly or indirectly, existential implications. Should we take the traditional connection to be severed in either cases? An argument to favor abstractive over axiomatic definitions on this score has been offered by neologicists by appealing to the notion of *arrogance*. Consider the following purported definition of the name 'Jack the Ripper,' as laid down at the beginning of the relevant police investigation:

> Jack the Ripper is the perpetrator of this series of killings. (JTR)

Are we warranted in taking JTR as true just by stipulating it? Not according to neologicists: to vindicate its truth we must ascertain both that those killings were actually perpetrated by someone (rather than being accidental events), and that the perpetrator was unique. Both presuppositions cannot be obtained merely by understanding the JTR statement itself. They are genuine pieces of (empirical) knowledge that must be independently obtained. Hence, JTR "cannot justifiably be affirmed without collateral (*a posteriori*) epistemic work" (Hale and Wright 2000, 128), so that "the truth of the vehicle of the stipulation is hostage to the obtaining of conditions of which it's reasonable to demand an independent assurance" (Hale and Wright 2009a, 465). Hence, JTR is an arrogant definition, while "A good implicit definition has to be something which we can freely stipulate as true, without any additional epistemological obligation" (Hale and Wright 2000, 133). Conversely, HP is an innocent definition: its truth can be appreciated without any further epistemological obligations: HP alone neither entails (see Section 5.2.2), nor presupposes, for its truth, the truth of any existential claim (but see MacFarlane 2009, Assadian 2023 for contrasting thoughts).

Arrogance becomes crucial in distinguishing the epistemic pedigree of HP vis-à-vis the stipulation of the PA^2 Axioms, and hence in comparing abstractionism and structuralism. Stipulating the PA^2 Axioms (given the occurrence of simple predications and existential claims), is tantamount to stipulating that a model for them (a progression of suitable objects) exists. But no mere stipulation can guarantee existence in this way, and ascertaining that the axioms are true would thus require "collateral [...] epistemic work" (be it *a posteriori* or *a priori*). The stipulation of the axioms is therefore arrogant and cannot sustain *a priori* knowledge.

A full analysis of arrogance is still lacking. It is a complex and promising notion, which, if stable, can distinguish between licit and illicit definitions by also accounting for some of the requirements traditionally imposed by LOGICALITY and ANALYTICITY, without endorsing those requirements themselves – unless, of course, arrogance is just analyticity in sheep's clothing.

7.6.2 Applicability and Frege's Constraint

Foundational programs traditionally focus on pure mathematics, and our discussion only cursorily mentioned applied mathematics. A recent debate bridges the two, and leads to a final comparison between axiomatic and abstractive definitions.

We saw that Frege rejects bringing particular applications into the content of mathematical statements. However, he believed that *applicability*, in full generality, must be accounted for. This does not entail that a definition of a mathematical notion isn't justified if we don't have a particular application in mind (famously, e.g., group theory was developed in complete autonomy from empirical applications), but it does entail that we should at least in principle understand how the notion could be applied, on pain of introducing a purely symbolic apparatus. Indeed, in his critique of formalist views like Thomae's and Heine's, in order to distinguish *contentual* (*inhaltliche*) arithmetic from a chess-like manipulation of language, he argues (Frege 1893–1903, II, §91) that "it is applicability alone which elevates arithmetic from a game to the rank of a science." For these reasons, Frege endorsed an additional criterion for his definitions, which Wright 2000 (p. 324) expresses as follows:

FREGE'S CONSTRAINT (FC). A satisfactory foundation for a mathematical theory must somehow build its applications, actual and potential, into its core – into the content it ascribes to the statements of the theory – rather than merely "patch them on from the outside" [The quotation is from Frege 1893–1903, II, §159.].

Apparently, HP meets the constraint. If canonical applications of cardinals are grounded on comparison of cardinality between concepts, the RHS of HP encodes exactly this kind of explanation of cardinal applications. But Wright endorses a robust reading of FC, according to which for abstractions to meet

FC requires viewing "the applications of the sorts of mathematical objects they concern as belonging to the essence of objects of those sorts" (Wright 2000, 325). The emphasis on essences coheres with a possible interpretation of Frege's views as substantially concerned with the nature of arithmetical objects. It also furnishes another argument against structuralist definitions by entailing that arithmetical objects "have an essence which transcends whatever is shared by the respective types of models of even categorical (second-order) formulations of those theories" (Wright 2000, 325). Hence no axiomatic, structural definition of natural numbers can meet FC. This objection is strictly related to that of arrogance. Hume's Principle conveys the idea that the essence of cardinals is tied to their applications, whereas the PA^2 Axioms "convey no more than the collective structure of the finite cardinals" (Hale and Wright 2009a, 471): any application, through appropriate mappings between target domains and the structure (or part of it), "will then depend upon an additional appreciation of structural affinities between any such instance and the intended realm of application" (Wright 2000, 326–327). Structural definitions are arrogant (also) because they fail to meet FC.

This can be questioned in various ways. The robust reading of FC seems almost question-begging against structuralists. Once weakened, FC may be satisfiable by *ante rem* structuralists too (Sereni 2019). After all, structure concepts will be defined (also) in terms of isomorphism. Since their applications are also based on (iso)morphisms, the resources for explaining applicability will be built into, and not additional to, their (axiomatic) definition.

More generally, Wright's argument relies on the assumption that canonical arithmetical applications are cardinal, rather than ordinal, in nature. But not only what is to count as canonical can depend on various theoretical purposes (Panza and Sereni 2019). Also, structuralists may retort that either view must account for both cardinal and ordinal applications, and to do so can appeal to available resources, although apparently additional to their respective definitions. So either both fail FC, or both meet it (Snyder, Samuels, and Shapiro 2018).

The emphasis on applications introduces a novel element in the comparison of rival mathematical definitions, both for arithmetic and analysis. Indeed, FC seems to underlie Frege's objections to Cantor's and Dedekind's definitions of the reals, on grounds that in them "either measurement does not feature at all, or […] it features without any internal connection grounded in the nature of the number itself, but is merely tacked on externally" (Frege 1893–1903, §159; see also Dummett 1991, 272–273). While Hale (2000) concurs that FC should be met by an abstractive definition of reals, Wright (2000) believes FC is motivated only when "the flow of concept formation" (329) can go from applications to theory, so that at least some rudimentary conception of the relevant numbers can be conveyed via elementary applications, something that cannot happen with the reals, given limitations in the exactness of empirical measurements.

As some reactions have displayed (Shapiro 2000, 361; Batitsky 2002), there are both philosophical and practical reasons to dismiss this constraint. A preeminent regard for structural relations, as well as the acknowledgment of how applications proceed in actual mathematics (often by taking number systems for granted and applying suitable representation theorems), may lead to dismissing constraints on definitions which are perceived as either overrationalistic or detached from mathematical practice. On the other hand, autonomy can still be advocated for epistemological and foundational inquiries.

***

As this last debate emphasizes, the traditional relevance of mathematical definitions (as well as elucidations, and explications), which we have tried to address in our discussion, is primarily due to the peculiar spot they occupy, in between practice and theory. They are pillars around which different needs revolve: the need to understand and recover a multiplicity of heterogeneous ordinary and scientific activities; and the need to systematize and formalize mathematical theories in a way that duly answers to genuinely philosophical concerns and desiderata. For these reasons, clarifying their epistemological import is still among the most pressing matters in the philosophy of mathematics.

References

Antonelli, A. (1998). Definition. In *Routledge Encyclopedia of Philosophy* (pp. 150–154). London: Routledge. DOI: https://doi.org/10.4324/9780415249126-Y057-1.

Antonelli, A., & May, R. C. (2000). Frege's New Science. *Notre Dame Journal of Formal Logic, 41*(3), 242–270.

Antos, C., Friedman, S., Honzik, R., & Ternullo, C. (Eds.). (2018). *The Hyperuniverse Project and Maximality*. Basel: Birkhäuser.

Aristotle. (1984). *The Complete Works of Aristotle* (Vol. 1; J. Barnes). Princeton, NJ: Princeton University Press.

Assadian, B. (2023). Abstraction and Semantic Presuppositions. *Analysis, 15*(3), 419–428.

Ayer, A. J. (1936). *Language, Truth and Logic*. London: V. Gollancz.

Balaguer, M. (1998). *Platonism and Anti-Platonism in Mathematics*. York, NY: Oxford University Press.

Barton, N. (2024). *Iterative Conceptions of Set*. Cambridge: Cambridge University Press.

Batitsky, V. (2002). Some Measurement-Theoretic Concerns about Hale's "Reals by Abstraction." *Philosophia Mathematica, 10*(3), 286–303.

Beall, J., & Restall, G. (2005). *Logical Pluralism*. Oxford: Oxford University Press.

Beaney, M. (2021). Analysis. In E. N. Zalta (Ed.), *The Stanford Encyclopedia of Philosophy* (Summer 2021 ed.). https://plato.stanford.edu/archIves/spr2010/entries/frege-hilbert/index.html.

Belnap, N. (1993). On Rigorous Definitions. *Philosophical Studies, 72*(2–3), 115–146.

Benacerraf, P. (1965). What Numbers Could Not Be. *Philosophical Review, 74*(1), 47–73.

Benacerraf, P. (1973). Mathematical Truth. *Journal of Philosophy, 70*(19), 661–679.

Benacerraf, P. (1981). Frege: The Last Logicist. *Midwest Studies in Philosophy, 6*(1), 17–36.

Benacerraf, P. (1996). What Mathematical Truth Could Not Be – 1. In A. Morton & S. P. Stich (Eds.), *Benacerraf and His Critics* (pp. 9–59). Oxford: Clarendon Press.

Bentham, J. (1962). A Fragment on Ontology. In J. Bowring (Ed.), *The Works of Jeremy Bentham* (Vol. VIII, pp. 193–211). New York, NY: Russell & Russell Inc.

Bentzen, B. (2019). Frege on Referentiality and Julius Caesar in Grundgesetze Section 10. *Notre Dame Journal of Formal Logic, 60*(4), 617–637.

Beth, E. (1953). On Padoa's Method in the Theory of Definition. *Indagationes Mathematicae, 15*, 330–339.

Biagioli, F. (2023). Federigo Enriques and the Philosophical Background to the Discussion of Implicit Definitions. In P. Cantù & G. Schiemer (Eds.), *Logic, Epistemology, and Scientific Theories? From Peano to the Vienna Circle* (pp. 153–174). Cham: Springer Nature.

Blanchette, P. A. (1994). Frege's Reduction. *History and Philosophy of Logic. 15*(1), 85–103.

Blanchette, P. A. (2007). Frege on Consistency and Conceptual Analysis. *Philosophia Mathematica, 15*(3), 321–346.

Blanchette, P. A. (2018). The Frege–Hilbert Controversy. In E. N. Zalta (Ed.), *The Stanford Encyclopedia of Philosophy* (Fall 2018 ed.). https://plato .stanford.edu/archIves/spr2010/entries/frege-hilbert/index.html.

Boccuni, F., & Panza, M. (2022). Frege's Theory of Real Numbers: A Consistent Rendering. *The Review of Symbolic Logic, 15*(3), 624–667.

Boccuni, F., & Sereni, A. (Eds.). (2021). *Origins and Varieties of Logicism: On the Logico-Philosophical Foundations of Logicism.* Abingdon: Routledge.

Boccuni, F., & Sereni, A. (in press). The Logics of Abstraction. In F. Ferrari, M. Carrara, O. Hjortland, G. Sher, G. Sagi, and E. Brendel (Eds.), *Oxford Handbook of Philosophy of Mathematics and Logic.* Oxford: Oxford University Press.

Boccuni, F., & Woods, J. (2020). Structuralist Neologicism. *Philosophia Mathematica, 28*(3), 296–316.

Boccuni, F., & Zanetti, L. (in press). *Abstractionism.* Cambridge: Cambridge University Press.

Boddy, R. (2021). Frege on the Fruitfulness of Definitions. *Journal for the History of Analytical Philosophy, 9*(11), 99–114.

Boghossian, P. A. (1996a). Analyticity. In B. Hale & C. Wright (Eds.), *A Companion to the Philosophy of Language* (pp. 331–368). Hoboken, NJ: Wiley-Blackwell.

Boghossian, P. A. (1996b). Analyticity Reconsidered. *Noûs, 30*(3), 360–391.

Boolos, G. (1986). Saving Frege from Contradiction. *Proceedings of theAristotelian Society*, 87 (1986/87): 137–151.

Boolos, G. (1987). The Consistency of Frege's Foundations of Arithmetic. In J. Thomson (Ed.), *On Being and Saying: Essays in Honor of Richard Cartwright* (pp. 3–20). Cambridge, MA: Massachusetts Institute of Technology Press.

Boolos, G. (1997). Is Hume's Principle Analytic? In R. K. Heck (Ed.), *Language, Thought and Logic* (pp. 245–261). Oxford: Clarendon Press (this and subsequent work originally published under the name "Richard G. Heck, Jr").

Boolos, G. S., Burgess, J. P., & Jeffrey, R. C. (2007). *Computability and Logic* (5 ed.). Cambridge, UK: Cambridge University Press.

Bourbaki, N. (1950). The Architecture of Mathematics. *The American Mathematical Monthly, 57*(4), 221–232.

Breckenridge, W., & Magidor, O. (2012). Arbitrary Reference. *Philosophical Studies, 158*(3), 377–400.

Brown, J. R. (2008). *Philosophy of Mathematics: A Contemporary Introduction to the World of Proofs and Pictures*. New York, NY: Routledge.

Brun, G. (2016). Explication as a Method of Conceptual Re-Engineering. *Erkenntnis, 81*(6), 1211–1241.

Brun, G. (2020). Conceptual Re-Engineering: From Explication to Reflective Equilibrium. *Synthese, 197*(3), 925–954.

Burali-Forti, C. (1894). *Logica Matematica*. Milan: Ulrico Hoepli.

Burge, T. (1984). Frege on Extensions of Concepts, from 1884 to 1903. *Philosophical Review, 93*(1), 3–34.

Burgess, J. P. (2005). *Fixing Frege*. Princeton, NJ: Princeton University Press.

Burgess, J. P. (2015). *Rigor and Structure*. Oxford: Oxford University Press.

Burgess, J. P., & Rosen, G. (1997). *A Subject with No Object*. New York, NY: Oxford University Press.

Cantor, G. (1872). Über die Ausdehnung eines Satzes aus der Theorie der trigonometrischen Reichen. *Mathematische Annalen, 5*, 123–132.

Cantù, P. (2023). What Is Axiomatics? *Annals of Mathematics and Philosophy, 1*, 1–24.

Cantù, P., & Luciano, E. (2021). Giuseppe Peano and His School: Axiomatics, Symbolism and Rigor. *Philosophia Scientiæ, 25*, 3–14. https://link.springer.com/collections/difhbeaaed

Cantù, P., & Testa, I. (Eds.). (2023). *Mathematical Practice and Social Ontology* (Vol. 42). *Topoi*, special issue. https://link.springer.com/collections/difhbeaaed.

Carey, S. (2009). *The Origin of Concepts*. Oxford: Oxford University Press.

Carnap, R. (1928). *Der Logische Aufbau der Welt*. Hamburg: Meiner Verlag. (Eng. trans. *The Logical Structure of the World*, by R. A. George, Berkeley, CA: University of California Press, 1967 ed.).

Carnap, R. (1947). *Meaning and Necessity: A Study in Semantics and Modal Logic*. Chicago, IL: University of Chicago Press.

Carnap, R. (1950a). Empiricism, Semantics and Ontology. *Revue Internationale de Philosophie, 4*(11), 20–40.

Carnap, R. (1950b). *Logical Foundations of Probability*. Chicago, IL: University of Chicago Press.

Carrara, M., & Zanetti, L. (2023). *Thin Objects* (Vol. 89) (No. 3). *Theoria*, special issue. https://onlinelibrary.wiley.com/doi/10.1111/theo.12463.

Coffa, A. (1991). *The Semantic Tradition From Kant to Carnap: To the Vienna Station*. New York, NY: Cambridge University Press.

Cole, J. C. (2015). Social Construction, Mathematics, and the Collective Imposition of Function onto Reality. *Erkenntnis, 80*(6), 1101–1124.

Cook, R. T. (2007). *The Arché Papers on the Mathematics of Abstraction*. Dordrecht: Springer.

Cook, Roy T. (2012). Conservativeness, Stability, and Abstraction. *British Journal for the Philosophy of Science 63*(3), 673–696.

Cook, R. T. (2021). Logicism, Separation and Complement. In F. Boccuni & A. Sereni (Eds.), *Origins and Varieties of Logicism* (pp. 289–308). Abingdon: Routledge.

Cook, R. T. (2023). Frege's Logic. In E. N. Zalta & U. Nodelman (Eds.), *The Stanford Encyclopedia of Philosophy* (Spring 2023 ed.). https://plato.stanford.edu/entries/frege-logic/.

Cook, R. T., & Linnebo, O. (2018). Cardinality and acceptable abstraction. *Notre Dame Journal of Formal Logic, 59*(1), 61–74.

Coumans, V. J. W. (2024). Definitions (and Concepts) in Mathematical Practice. In B. Sriraman (Ed.), *Handbook of the History and Philosophy of Mathematical Practice* (pp. 135–157). Cham: Springer.

De Toffoli, S. (2021). Groundwork for a Fallibilist Account of Mathematics. *Philosophical Quarterly, 7*(4), 823–844.

Decock, L. (2022). Frege's Theorem and Mathematical Cognition. In F. Boccuni & A. Sereni (Eds.), *Origins and Varieties of Logicism: On the Logicophilosophical Foundations of Logicism* (pp. 372–394). New York, NY: Routledge.

Dedekind, R. (1872). *Stetigkeit und irrationale Zahlen*. Braunschweig: Vieweg.

Dedekind, R. (1888). *Was sind und was sollen die Zahlen?* Braunschweig: Vieweg (Trans. in *Essays on the Theory of Numbers*, The Open Court Publishing Company, Chicago, IL, 1901, pp. 31–115).

Dehaene, S. (2011). *The Number Sense: How the Mind Creates Mathematics*. New York, NY: Oxford University Press.

de Jong, W. R., & Betti, A. (2010). The Classical Model of Science: A Millennia-Old Model of Scientific Rationality. *Synthese, 174*(2), 185–203.

deRosset, L., & Linnebo, O. (2023). Abstraction and grounding. *Philosophy and Phenomenological Research, 109*(1), 357–390.

Descartes, R. (1637). Discourse and Essays. In J. Cottingham, R. Stoothoff, & D. Murdoch (Eds.), *The Philosophical Writings of Descartes, 1985* (Vol. 1, pp. 109–176). Cambridge: Cambridge University Press.

Deslauriers, M. (2007). *Aristotle on Definition*. Boston, MA: Brill.

Donaldson, T. (2017). The (Metaphysical) Foundations of Arithmetic? *Noûs, 51*(2), 775–801.

Dubislav, W. (1981). *Die Definition*. Hamburg: Meiner.

Dummett, M. (1973). *Frege: Philosophy of Language*. London: Duckworth.

Dummett, M. (1991). *Frege: Philosophy of Mathematics*. Cambridge, MA: Harvard University Press.

Dummett, M. (1993a). Discussions: Chairman's Address: Basic Law V. *Proceedings of the Aristotelian Society, 94*, 243–252.

Dummett, M. (1993b). *Origins of Analytical Philosophy*. Cambridge, MA: Harvard University Press.

Dummett, M. (1993c). *The Seas of Language*. New York, NY: Oxford University Press.

Dummett, M. (1998). Neo-Fregeans: In Bad Company? In M. Schirn (Ed.), *The Philosophy of Mathematics Today*. Oxford: Clarendon Press.

Ebert, P. A. (2016). A Framework for Implicit Definitions and the A Priori. In P. A. Ebert & M. Rossberg (Eds.), *Abstractionism* (pp. 133–160). Oxford: Oxford University Press.

Ebert, P. A., & Rossberg, M. (2016). *Abstractionism*. Oxford: Oxford University Press.

Ebert, P. A., & Rossberg, M. (2019). Mathematical Creation in Frege's Grundgesetze. In P. A. Ebert & M. Rossberg (Eds.), *Essays on Frege's Basic Laws of Arithmetic* (pp. 325–342). Oxford: Oxford University Press.

Euclid. (1926). *The Thirteen Books of the Elements*. Cambridge: Cambridge University Press. (Translated with introduction and commentary by Sir Thomas L. Heath; 3 vols.)

Falguera, J. L., Martínez-Vidal, C., & Rosen, G. (2022). Abstract Objects. In E. N. Zalta (Ed.), *The Stanford Encyclopedia of Philosophy* (Summer 2022 ed.). https://plato.stanford.edu/entries/abstract-objects/.

Ferreirós, J. (2007). *Labyrinth of Thought: A History of Set Theory and Its Role in Modern Mathematics*. (2nd ed.). Basel: Birkäuser.

Ferreirós, J. (2023). Conceptual Structuralism. *Journal for General Philosophy of Science, 54*(1), 125–148.

Field, H. (1974). Quine and the Correspondence Theory. *Philosophical Review, 83*(2), 200–228.

Field, H. (1980/2016). *Science without Numbers*. Oxford: Basil Blackwell.

Field, H. (1989). *Realism, Mathematics and Modality*. Oxford: Blackwell.

Fine, K. (2002). *The Limits of Abstraction*. Oxford: Oxford University Press.

Fine, K. (2012). Guide to Ground. In F. Correia & B. Schnieder (Eds.), *Metaphysical Grounding* (pp. 37–80). Cambridge: Cambridge University Press.

Frans, J., Coumans, V., & de Regt, H. (2022). Explanation, Understanding, and Definitions in Mathematics. *Logique et Analyse, 257*, 79–99.

Frege, G. (1879). *Begriffsschrift, eine der Arithmetischen nachgebildete Formelsprache des reinen Denkens*. Halle: Nebert. English translation in van Heijenoort (1967).

Frege, G. (1884). *Die Grundlagen der Arithmetik: eine logische mathematische Untersuchung über den Begriff der Zahl*. Breslau: Koebner. English translation by J. Austin, *The Foundations of Arithmetic: A Logico-Mathematical Enquiry into the Concept of Number*, Blackwell, Oxford, 1950/1953.

Frege, G. (1892). Über Sinn und Bedeutung. *Zeitschrift für Philosophie und philosophische Kritik, Vol. 100*, 25–50.

Frege, G. (1893–1903). *Die Grundgesetze der Arithmetick* (Vol. I–II). Jena: H. Phole. English translation by P. Ebert, M Rossberg, ed. C. Wright, *Basic Laws of Arithmetic*, Oxford University Press, Oxford, 2016.

Frege, G. (1971). *On the Foundations of Geometry and Formal Theories of Arithmetic* (E.-H. W. Kluge, Ed.). New Haven, CT: Yale University Press.

Frege, G. (1979a). Logic in Mathematics. In *Posthumous Writings*. Oxford: Blackwell.

Frege, G. (1979b). *Posthumous Writings*. Oxford: Blackwell.

Frege, G. (1980). *Philosophical and Mathematical Correspondence*. Oxford: Basil Blackwell.

Frege, G. (1984). *Collected Papers on Mathematics, Logic and Philosophy*. Oxford: Basil Blackwell.

Frege, G., & Carnap, R. (2003). *Frege's Lectures on Logic: Carnap's Student Notes, 1910–1914* (E. H. Reck and S. Awodey, Eds.). Chicago, IL: Open Court.

Gabriel, G. (1978). Implizite Definitionen: eine Verwechselungsgeschichte. *Annals of Science, 35*(4), 419–423.

Gergonne, J. D. (1818–9). Essai sur la théorie des definitions. *Annales de Mathématiques Pures et Appliquée, 9*, 1–35.

Giaquinto, M. (2007). *Visual Thinking in Mathematics*. New York: Oxford University Press.

Gideon Rosen, S. Y. (2020). Solving the Caesar Problem – with Metaphysics. In *Logic, Language, and Mathematics: Themes from the Philosophy of Crispin Wright* (pp. 116–132). Oxford: Oxford University Press.

Giovannini, E. N., & Schiemer, G. (2019). What Are Implicit Definitions? *Erkenntnis, 86*(6), 1661–1691.

Gödel, K. (1964 (1st ed. 1947)). What Is Cantor's Continuum Problem (1964 Version). *Journal of Symbolic Logic,* (2), 116–117.

Goldman, A. I. (1967). A Causal Theory of Knowing. *Journal of Philosophy, 64*(12), 357–372.

Goodman, N. (1954/1983). *Fact, Fiction, and Forecast* (4th ed.). Cambridge, MA: Harvard University Press.

Graham, P., & Pedersen, N. J. L. L. (Eds.). (2020). *Epistemic Entitlement.* Oxford: Oxford University Press.

Gupta, A., & Mackereth, S. (2023). Definitions. In E. N. Zalta & U. Nodelman (Eds.), *The Stanford Encyclopedia of Philosophy* (Fall 2023 ed.). https://plato.stanford.edu/entries/definitions/.

Hale, B. (1988). *Abstract Objects.* New York, NY: Blackwell.

Hale, B. (1994). Dummett's Critique of Wright's Attempt to Resuscitate Frege. In B. Hale & C. Wright (Eds.), *Reason's Proper Study* (pp. 189–213). Oxford: Oxford University Press.

Hale, B. (1997). Grundlagen §64. *Proceedings of the Aristotelian Society, 97*(3), 243–261.

Hale, B. (2000). Reals by Abstraction. *Philosophia Mathematica (III), 8,* 100–123.

Hale, B. (2013). *Necessary Beings.* Oxford: Oxford University Press.

Hale, B., & Wright, C. (2000). Implicit Definition and the A Priori. In P. Boghossian & C. Peacocke (Eds.), *New Essays on the A Priori* (pp. 286–319). Oxford: Oxford University Press.

Hale, B., & Wright, C. (2001a). *Reason's Proper Study: Essays Towards a Neo-Fregean Philosophy of Mathematics.* Oxford: Oxford University Press.

Hale, B., & Wright, C. (2001b). To Bury Caesar … In *The Reason's Proper Study* (pp. 335–396). New York, NY: Oxford University Press.

Hale, B., & Wright, C. (2002). Benacerraf's Dilemma Revisited. *European Journal of Philosophy, 10*(1), 101–8211.

Hale, B., & Wright, C. (2008). Abstraction and Additional Nature. *Philosophia Mathematica, 16*(2), 182–208.

Hale, B., & Wright, C. (2009a). Focus Restored: Comments on John MacFarlane. *Synthese, 170*(3), 457–482.

Hale, B., & Wright, C. (2009b). The Metaontology of Abstraction. In D. Chalmers, D. Manley, & R. Wasserman (Eds.), *Metametaphysics* (pp. 178–212). Oxford: Oxford University Press.

Hallett, M. (2019). Frege on Creation. In P. A. Ebert & M. Rossberg (Eds.), *Essays on Frege's Basic Laws of Arithmetic* (pp. 285–324). Oxford: Oxford University Press.

Hallett, M. (2021). Frege and Hilbert on Conceptual Analysis and Foundations. In F. Boccuni & A. Sereni (Eds.), *Origins and Varieties of Logicism*. Abingdon: Routledge.

Hamkins, J. D. (2012). The Set-Theoretic Multiverse. *The Review of Symbolic Logic, 5*(3), 416–449.

Heck, R. K. (1993). The Development of Arithmetic in Frege's *Grundgesetze der Arithmetik*. *Journal of Symbolic Logic, 58*(2), 579–601.

Heck, R. K. (1999). Grundgesetze der Arithmetic I §10. *Philosophia Mathematica, 7*(3), 258–292.

Heck, R. K. (2000). Cardinality, Counting, and Equinumerosity. *Notre Dame Journal of Formal Logic, 41*(3), 187–209.

Heck, R. K. (2011). *Frege's Theorem*. Oxford: Oxford University Press.

Hellman, G., & Shapiro, S. (2018). *Mathematical Structuralism*. Cambridge: Cambridge University Press.

Hilbert, D. (1899). *Die Grundlagen der Geometrie*. Leipzig: Teubner. English translation in Hilbert, D. *The Foundations of Geometry*. Chicago: Open Court, 1902.

Hilbert, D. (1900). On the Concept of Number. In W. B. Ewald (Ed.), *From Kant to Hilbert: A Source Book in the Foundations of Mathematics (1996)* (pp. 1089–1095). Oxford: Oxford University Press.

Hilbert, D. (1926). Uber das Unendliche. *Mathematische Annalen, 95,* 161–190.

Hodges, W. (1993). *Model Theory*. Cambridge: Cambridge University Press.

Hodges, W. (2023). Model Theory. In E. N. Zalta & U. Nodelman (Eds.), *The Stanford Encyclopedia of Philosophy* (Fall 2023 ed.). https://plato.stanford.edu/entries/model-theory/.

Hume, D. (1739–40). *A Treatise of Human Nature*. London: John Noon (Books I–II) and Thomas Longman (Book III). (3 vols.).

Hume, D. (1748). *An Enquiry Concerning Human Understanding*. London: A. Millar.

Jeshion, R. (2001). Frege's Notions of Self-Evidence. *Mind, 110*(440), 937–976.

Jeshion, R. (2004). Frege: Evidence for Self-Evidence. *Mind, 113*(449), 131–138.

Kant, I. (1781). *Critique of Pure Reason*. Cambridge: Cambridge University Press.

Kitcher, P. (1983). *The Nature of Mathematical Knowledge*. Oxford: Oxford University Press.

Korbmacher, J., & Schiemer, G. (2018). What Are Structural Properties? *Philosophia Mathematica, 26*(3), 295–323.

Lakatos, I. (Ed.). (1976). *Proofs and Refutations*. New York, NY: Cambridge University Press.

Landry, E. (2013). The Genetic versus the Axiomatic Method: Responding to Feferman 1977. *Review of Symbolic Logic, 6*(1), 24–51.

Leach-Krouse, G. (2017). Structural-Abstraction Principles. *Philosophia Mathematica, 25*(1), 45–72.

Lésniewski, S. (1931). Über Definitionen in der sogenannten Theorie der Deduktion. *Comptes Rendus des Séances de la Société des Sciences et des Lettres de Varsovie (Classe 3), 24.*

Liggins, D. (2010). Epistemological Objections to Platonism. *Philosophy Compass, 5*(1), 67–77.

Linnebo, Ø. (2003). Frege's Conception of Logic: From Kant to *Grundgesetze. Manuscrito, 26*(2), 235–252.

Linnebo, Ø. (2004). Frege's Proof of Referentiality. *Notre Dame Journal of Formal Logic, 45*(2), 73–98.

Linnebo, Ø. (2006). Epistemological Challenges to Mathematical Platonism. *Philosophical Studies, 129*(3), 545–574.

Linnebo, Ø. (2008). Structuralism and the Notion of Dependence. *Philosophical Quarterly, 58*(230), 59–79.

Linnebo, Ø. (Ed.). (2009). *The Bad Company Problem. Synthese 170*(3).

Linnebo, Ø. (2016). Impredicativity in the Neo-Fregean Program. In P. A. Ebert & M. Rossberg (Eds.), *Abstractionism* (pp. 247–268). Oxford: Oxford University Press.

Linnebo, Ø. (2017). *Philosophy of Mathematics*. Princeton, NJ: Princeton University Press.

Linnebo, Ø. (2018a). Dummett on Indefinite Extensibility. *Philosophical Issues, 28*(1), 196–220.

Linnebo, Ø. (2018b). *Thin Objects*. Oxford: Oxford University Press.

Linnebo, Ø. (2022). Plural Quantification. In E. N. Zalta (Ed.), *The Stanford Encyclopedia of Philosophy* (Spring 2022 ed.). https://plato.stanford.edu/entries/plural-quant/.

Linnebo, Ø. (2023). Platonism in the Philosophy of Mathematics. In E. N. Zalta & U. Nodelman (Eds.), *The Stanford Encyclopedia of Philosophy* (Summer 2023 ed.). https://plato.stanford.edu/entries/platonism-mathematics/.

Linnebo, Ø., & Pettigrew, R. (2014). Two Types of Abstraction for Structuralism. *Philosophical Quarterly, 64*(255), 267–283.

Locke, J. (1690). *An Essay Concerning Human Understanding* (P. H. Nidditch, Ed.). Oxford: Oxford University Press.

MacFarlane, J. (2002). Frege, Kant, and the Logic in Logicism. *The Philosophical Review*, *111*(1), 25–65.

MacFarlane, J. (2009). Double Vision: Two Questions about the Neo-Fregean Program. *Synthese*, *170*(3), 443–456.

Mackereth, S. (in press). Neologicism and Conservativeness. *Journal of Philosophy*. https://philpapers.org/versions/MACNAC-6

Maddy, P. (2011). *Defending the Axioms: On the Philosophical Foundations of Set Theory*. Oxford: Oxford University Press.

Mancosu, P. (2016). *Abstraction and Infinity*. Oxford: Oxford University Press.

Manders, K. (2008). The Euclidean Diagram. In P. Mancosu (Ed.), *The Philosophy of Mathematical Practice* (pp. 80–133). Oxford: Oxford University Press.

Mates, B. (1972). *Elementary Logic* (2nd ed.). New York, NY: Oxford University Press.

May, R. C., & Wehmeier, K. F. (2019). The Proof of Hume's Principle. In *Essays on Frege's Basic Laws of Arithmetic*. Oxford: Oxford University Press.

McGee, V. (1997). How We Learn Mathematical Language. *Philosophical Review*, *106*(1), 35–68.

McKenzie, K. (2022). *Fundamentality and Grounding*. Cambridge: Cambridge University Press.

McLarty, C. (1993). Numbers Can Be Just What They Have to. *Noûs*, *27*(4), 487–498.

Menger, K. (1943). What Is Dimension? *The American Mathematical Monthly*, *50*(1), 2–7.

Mill, J. S. (1843). *A System of Logic, Ratiocinative and Inductive*. Cambridge: Cambridge University Press.

Mueller, I. (1981). *Philosophy of Mathematics and Deductive Structure in Euclid's Elements*. Cambridge, MA: Massachusetts Institute of Technology Press.

Nodelman, U., & Zalta, E. N. (2024). Number Theory and Infinity Without Mathematics. *Journal of Philosophical Logic*, DOI: https://doi.org/10.1007/s10992-024-09762-7.

Novaes, C. D., & Reck, E. H. (2017). Carnapian Explication, Formalisms as Cognitive Tools, and the Paradox of Adequate Formalization. *Synthese*, *194*(1), 195–215.

Nutting, E. S. (2018). The Limits of Reconstructive Neologicist Epistemology. *Philosophical Quarterly*, *68*(273), 717–738.

Nutting, E. S. (2024). The Benacerraf Problem of Mathematical Truth and Knowledge. *The Internet Encyclopedia of Philosophy*, https://iep.utm.edu.

Otero, M. H. (1970). Gergonne on Implicit Definition. *Philosophy and Phenomenological Research*, *30*(4), 596–599.

Otte, M., & Panza, M. (Eds.). (1997). *Analysis and Synthesis in Mathematics*. Dordrecht: Kluwer Academic Publishers.

Padoa, A. (1900). Logical Introduction to Any Deductive Theory. In *From Frege to Gödel: A Source Book in Mathematical Logic, 1879–1931*. Cambridge, MA: Harvard University Press.

Pantsar, M. (2024). *Numerical Cognition and the Epistemology of Arithmetic*. Cambridge: Cambridge University Press.

Panza, M. (2016). *Abstraction and Epistemic Economy*. In Costreie, S. (ed.) *Early Analytic Philosophy – New Perspectives on the Tradition*. The Western Ontario Series in Philosophy of Science, vol. 80. Cham: Springer. DOI: https://doi.org/10.1007/978-3-319-24214-9_17.

Panza, M., & Sereni, A. (2013). *Plato's Problem. An Introduction to Mathematical Platonism*. London: Palgrave Macmillan.

Panza, M., & Sereni, A. (2019). Frege's Constraint and the Nature of Frege's Logicism. *Review of Symbolic Logic*, *12*(1), 97–143.

Panza, M., & Sereni, A. (in press). *The Other Frege*. Oxford: Oxford University Press.

Pascal, B. (1658). De l'Esprit géométrique et de l'Art de persuader. In *Oeuvres complètes*. Paris: Seuill.

Pasch, M. (1882). *Vorlesungen über neuere Geometrie* (2nd Edition 1926). Leipzig: Teubner.

Paseau, A. C., & Wrigley, W. (2024). *The Euclidean Programme*. Cambridge: Cambridge University Press.

Peano, G. (1889). *Arithmetices Principia, Nova Methodo Exposita* (in van Heijenohoort (1967), pp. 83–97). Turin: Fratelli Bocca.

Peano, G. (1895). *Formulaire de mathématiques*. Turin: Fratelli Bocca.

Pedersen, N. J. L. L. (2016). Hume's Principle and Entitlement: On the Epistemology of the Neo-Fregean Program. In P. Ebert & M. Rossberg (Eds.), *Abstractionism* (pp. 186–202). Oxford: Oxford University Press.

Picardi, E. (2022). Frege, Peano, and Russell on the Primitive Ideas of Logic. In A. Coliva (Ed.), *Frege on Language, Logic and Psychology* (pp. 101–130). New York, NY: Oxford University Press.

Pollard, S. (Ed.) (2010), *Essays on the Foundations of Mathematics by Moritz Pasch*, Dordrecht, Springer.

Posy, C. J. (2020). *Mathematical Intuitionism*. Cambridge: Cambridge University Press.

Potter, M. D. (2004). *Set Theory and Its Philosophy.* Oxford: Oxford University Press.

Potter, M. D., & Sullivan, P. (2005). What Is Wrong with Abstraction? *Philosophia Mathematica, 13*(2), 187–193.

Priest, G. (2024). *Mathematical Pluralism.* Cambridge: Cambridge University Press.

Quine, W. V. O. (1936). Truth by Convention. In A. N. Whitehead (Ed.), *Philosophical Essays for Alfred North Whitehead.* New York, NY: Longman, Green, & Company Inc.

Quine, W. V. O. (1951). Two Dogmas of Empiricism. *Philosophical Review, 60*(1), 20–43.

Quine, W. V. O. (1954). Carnap and Logical Truth. *Synthese, 12*(4), 350–374.

Quine, W. V. O. (1960). *Word and Object.* Cambridge, MA: Massachusetts Institute of Technology Press.

Quine, W. V. O. (1969). Epistemology Naturalized. In *Ontological Relativity and Other Essays.* New York, NY: Columbia University Press.

Quine, W. V. O. (1970). *Philosophy of Logic* (2nd Edition 1986). Cambridge, MA: Harvard University Press.

Raatikainen, P. (2022). Gödel's Incompleteness Theorems. In E. N. Zalta (Ed.), *The Stanford Encyclopedia of Philosophy* (Spring 2022 ed.). https://plato .stanford.edu/archives/spr2022/entries/goedel-incompleteness/.

Ramsey, F. P. (1931). Theories. In R. B. Braithwaite (Ed.), *The Foundations of Mathematics and Other Logical Essays.* London: Routledge & Kegan Paul.

Rayo, A. (2013). *The Construction of Logical Space.* Oxford: Oxford University Press.

Reck, E. H. (2007). Frege–Russell Numbers: Analysis or Explication? In M. Beaney (Ed.), *The Analytic Turn* (pp. 33–50). New York, NY: Routledge.

Reck, E. H. (2013). Frege, Dedekind, and the Origins of Logicism. *History and Philosophy of Logic, 34*(3), 242–265.

Reck, E. H. (2021). Dedekind's Logicism: A Reconsideration and Contextualization. In F. Boccuni & A. Sereni (Eds.), *Origins and Varieties of Logicism* (pp. 119–146). New York: Routledge.

Reck, E. H. (2024). Carnapian Explication: Origins and Shifting Goals. In A. W. Richardson & A. T. Tuboly (Eds.), *Interpreting Carnap* (pp. 127–149). Cambridge: Cambridge University Press.

Reck, E. H., & Price, M. P. (2000). Structures and Structuralism in Contemporary Philosophy of Mathematics. *Synthese, 125*(3), 341–383.

Reck, E. H., & Schiemer, G. (2023). Structuralism in the Philosophy of Mathematics. In E. N. Zalta (Ed.), *The Stanford Encyclopedia of Philosophy* (Spring 2023 ed.). https://plato.stanford.edu/entries/structuralism-mathematics/.

Robinson, R. (1950). *Definition*. Oxford: Clarendon Press.

Rosen, G. (2010). Metaphysical Dependence: Grounding and Reduction. In B. Hale & A. Hoffmann (Eds.), *Modality: Metaphysics, Logic, and Epistemology* (pp. 109–136). Oxford: Oxford University Press.

Russell, B. (1945). *History of Western Philosophy*. New York: Routledge.

Russo, L. (1998). The Definitions of Fundamental Geometric Entities Contained in Book I of Euclid's *Elements*. *Archive for History of Exact Sciences*, *52*(3), 195–219.

Samuels, R., & Snyder, E. (2024). *Number Concepts: An Interdisciplinary Inquiry*. Cambridge: Cambridge University Press.

Schwartzkopff, R. (2011). Numbers as Ontologically Dependent Objects: Hume's Principle Revisited. *Grazer Philosophische Studien*, *82*, 353–373.

Schwartzkopff, R. (2016). Singular Terms Revisited. *Synthese*, *193*(3), 909–936. DOI: https://doi.org/10.1007/s11229-015-0777-2.

Sereni, A. (2016). A Dilemma for Benacerraf's Dilemma? In F. Pataut (Ed.), *Truth, Objects, Infinity: New Perspectives on the Philosophy of Paul Benacerraf* (pp. 93–125). Cham: Springer.

Sereni, A. (2019). On the Philosophical Significance of Frege's Constraint. *Philosophia Mathematica*, *27*(2), 244–275.

Sereni, A., Sforza Fogliani, M. P., & Zanetti, L. (2023). For Better and for Worse: Abstractionism, Good Company, and Pluralism. *Review of Symbolic Logic*, *16*(1), 268–297.

Sereni, A., & Zanetti, L. (2023). Minimalism, Trivialism, Aristotelianism. *Theoria*, *89*(3), 280–297.

Shapiro, S. (1991). *Foundations without Foundationalism: A Case for Second-Order Logic*. New York, NY: Oxford University Press.

Shapiro, S. (1997). *Philosophy of Mathematics: Structure and Ontology*. Oxford: Oxford University Press.

Shapiro, S. (2000). Frege Meets Dedekind: A Neologicist Treatment of Real Analysis. *Notre Dame Journal of Formal Logic*, *4*, 317–421.

Shapiro, S. (2004). Foundations of Mathematics: Metaphysics, Epistemology, Structure. *The Philosophical Quarterly*, *54*, 16–37.

Shapiro, S. (2009). We Hold These Truths to Be Self-Evident: But What Do We Mean by That? *Review of Symbolic Logic*, *2*(1), 175–207.

Shapiro, S. (2011). Epistemology of Mathematics: What Are the Questions? What Count as Answers? *Philosophical Quarterly*, *61*(242), 130–150.

Shapiro, S. (2012). Objectivity, Explanation, and Cognitive Shortfall. In C. Wright & A. Coliva (Eds.), *Mind, Meaning, and Knowledge: Themes from the Philosophy of Crispin Wright*. Oxford: Oxford University Press.

Shapiro, S. (2014). *Varieties of Logic*. Oxford: Oxford University Press.

Shapiro, S., & Roberts, C. (2021). Open Texture and Mathematics. *Notre Dame Journal of Formal Logic*, *62*(1), 173–191.

Shapiro, S., & Wright, C. (2006). All Things Indefinitely Extensible. In A. Rayo & G. Uzquiano (Eds.), *Absolute Generality* (pp. 255–304). Oxford: Clarendon Press.

Shieh, S. (2008). Frege on Definitions. *Philosophy Compass*, *3*(5), 992–1012.

Simons, P. M. (1987). Frege's Theory of Real Numbers. *History and Philosophy of Logic*, *8*(1), 25–44. DOI: https://doi.org/10.1080/01445 348708837106.

Snyder, E., Samuels, R., & Shapiro, S. (2018). Neologicism, Frege's Constraint, and the Frege–Heck Condition. *Noûs*, *54*(1), 54–77.

Studd, J. P. (2016). Abstraction Reconceived. *British Journal for the Philosophy of Science*, *67*(2), 579–615. DOI: https://doi.org/10.1093/bjps/axu035.

Suppes, P. (1957). *Introduction to Logic*. Mineola, NY: Dover Publications.

Tanswell, F. S. (2018). Conceptual Engineering for Mathematical Concepts. *Inquiry: An Interdisciplinary Journal of Philosophy*, *61*(8), 881–913.

Tanswell, F. S. (2024). *Mathematical Rigour and Informal Proof*. Cambridge: Cambridge University Press.

Tappenden, J. (2008). Mathematical Concepts: Fruitfulness and Naturalness. In P. Mancosu (Ed.), *The Philosophy of Mathematical Practice* (pp. 276–301). Oxford: Oxford University Press.

Tappenden, J. (2012). Fruitfulness as a Theme in the Philosophy of Mathematics. *The Journal of Philosophy*, *109*(1/2), 204–219.

Tarski, A. (1933/1956). The Concept of Truth in Formalized Languages. In A. Tarski (Ed.), *Logic, Semantics, Metamathematics* (pp. 152–278). Oxford: Clarendon Press.

Tennant, N. (1987). *Anti-Realism and Logic: Truth as Eternal*. New York, NY: Oxford University Press.

Thomasson, A. L. (2014). *Ontology Made Easy*. New York, NY: Oxford University Press.

Šikić, Z. (2022). On Definitions in Mathematics. *Publications De L'Institut Mathémathique*, *112*(125), 41–52.

Uzquiano, G. (2015). Varieties of Indefinite Extensibility. *Notre Dame Journal of Formal Logic*, *56*(1), 147–166.

van Heijenoort, J. (Ed.). (1967). *From Frege to Gödel: A Source Book in Mathematical Logic, 1879–1931*. Cambridge, MA: Harvard University Press.

Waismann, F. (1945). Verifiability. *Proceedings of the Aristotelian Society, XIX,* 101–164.

Waismann, F. (1951). *Introduction to Mathematical Thinking.* Mineola, NY: Dover Publications.

Warren, J. (2020). *Shadows of Syntax.* New York, NY: Oxford University Press.

Weber, Z. (2022). *Paraconsistency in Mathematics.* Cambridge: Cambridge University Press.

Weiner, J. (1984). The Philosopher behind the Last Logicist. *Philosophical Quarterly, 34*(136), 242–264.

Weyl, H. (1949). *Philosophy of Mathematics and Natural Science* (O. Helmer, Ed.). Princeton, NJ: Princeton University Press.

White, N. P. (1974). What Numbers Are. *Synthese, 27*(1–2), 111–124.

Whitehead, A. N., & Russell, B. (1910–13). *Principia Mathematica.* Cambridge: Cambridge University Press. (3 vols.) Second edition 1925–27.

Wittgenstein, L. (1969). *On Certainty.* G. E. M. Anscombe and G. H. von Wright (eds.), New York, NY: Harper & Row.

Wright, C. (1983). *Frege's Conception of Numbers as Objects.* Aberdeen: Aberdeen University Press.

Wright, C. (1997). On the Philosophical Significance of Frege's Theorem. In R. K. Heck (Ed.), *Language, Thought, and Logic: Essays in Honour of Michael Dummett* (pp. 201–244). Oxford: Oxford University Press.

Wright, C. (1998). On the Harmless Impredicativity of N=('Hume's Principle'). In M. Schirn (Ed.), *The Philosophy of Mathematics Today* (pp. 339–368). Oxford: Clarendon Press.

Wright, C. (1999). Is Hume's Principle Analytic? *Notre Dame Journal of Formal Logic, 40*(1), 6–30.

Wright, C. (2000). Neo-Fregean Foundations for Real Analysis: Some Reflections on Frege's Constraint. *Notre Dame Journal of Formal Logic, 41*(4), 317–334.

Wright, C. (2004a). Intuition, Entitlement and the Epistemology of Logical Laws. *Dialectica, 58*(1), 155–175.

Wright, C. (2004b). Warrant for Nothing (and Foundations for Free)? *Aristotelian Society Supplementary Volume, 78*(1), 167–212.

Wright, C. (2007). On Quantifying into Predicate Position: Steps towards a New(Tralist) Perspective. In M. Leng, A. Paseau, & M. D. Potter (Eds.), *Mathematical Knowledge* (pp. 150–174). Oxford: Oxford University Press.

Wright, C. (2016). Abstraction and Epistemic Entitlement: On the Epistemological Status of Hume's Principle. In P. Ebert & M. Rossberg (Eds.), *Abstractionism* (pp. 161–185). Oxford: Oxford University Press.

Wright, C. (2020). Replies. In A. Miller (Ed.), *Logic, Language and Mathematics: Essays for Crispin Wright* (pp. 277–432). Oxford: Oxford University Press.

Yablo, S. (2005). The Myth of the Seven. In M. E. Kalderon (Ed.), *Fictionalism in Metaphysics* (pp. 88–115). Oxford: Clarendon Press.

Yap, A. (2014). Dedekind and Cassirer on Mathematical Concept Formation. *Philosophia Mathematica*, *25*(3), 369–389.

Zach, R. (2023). Hilbert's Program. In E. N. Zalta (Ed.), *The Stanford Encyclopedia of Philosophy* (Winter 2023 ed.). https://plato.stanford.edu/entries/hilbert-program/.

Zalta, E. N. (1999). Natural Numbers and Natural Cardinals as Abstract Objects: A Partial Reconstruction of Frege's *Grundgesetze* in Object Theory. *Journal of Philosophical Logic*, *28*(6), 619–60.

Zalta, E. N. (2023). Frege's Theorem and Foundations for Arithmetic. In E. N. Zalta (Ed.), *The Stanford Encyclopedia of Philosophy* (Summer 2023 ed.). https://plato.stanford.edu/entries/frege-theorem/.

Zalta, E. N. (1999). Natural Numbers and Natural Cardinals as Abstract Objects: A Partial Reconstruction of Frege's Grundgesetze in Object Theory. *Journal of Philosophical Logic*, *28*(6), 619–660.

Acknowledgments

I am indebted – through conversations, comments on previous drafts, bibliographical suggestions, or other varieties of occasional inspiration – to many colleagues, and especially: Bahram Assadian, Francesca Boccuni, Rachel Boddy, Nicolò Cambiaso, Ludovica Conti, Silvia De Toffoli, Eduardo Giovannini, Øystein Linnebo, Robert May, Marco Panza, Erich H. Reck, Colin Rittberg, Georg Schiemer, Crispin Wright, Luca Zanetti. I also wish to thank students who tested an earlier draft as class material (both at the School of Advanced Studies IUSS Pavia, at the University of Bergamo, and at San Raffaele University in Milan) as well as the PhD students in the HuME (*The Human Mind and its Explanations: Language, Brain and Reasoning*) PhD Program. My usual gratitude goes to my mother for bearing with so many sorry-I-have-to-work calls.

Research for this project has been supported by the MUR PRIN Italian National Project 2022TYNY32 – *Proof and understanding in mathematics. Purity of methods, simplicity, and explanation in mathematical reasoning* (funded by NextGeneration EU).

This is dedicated to Sarah, by definition, and to Luca, who doesn't know.

Cambridge Elements ≡

The Philosophy of Mathematics

Penelope Rush

University of Tasmania

From the time Penny Rush completed her thesis in the philosophy of mathematics (2005), she has worked continuously on themes around the realism/anti-realism divide and the nature of mathematics. Her edited collection *The Metaphysics of Logic* (Cambridge University Press, 2014), and forthcoming essay 'Metaphysical Optimism' (*Philosophy Supplement*), highlight a particular interest in the idea of reality itself and curiosity and respect as important philosophical methodologies.

Stewart Shapiro

The Ohio State University

Stewart Shapiro is the O'Donnell Professor of Philosophy at The Ohio State University, a Distinguished Visiting Professor at the University of Connecticut, and a Professorial Fellow at the University of Oslo. His major works include *Foundations without Foundationalism* (1991), *Philosophy of Mathematics: Structure and Ontology* (1997), *Vagueness in Context* (2006), and *Varieties of Logic* (2014). He has taught courses in logic, philosophy of mathematics, metaphysics, epistemology, philosophy of religion, Jewish philosophy, social and political philosophy, and medical ethics.

About the Series

This Cambridge Elements series provides an extensive overview of the philosophy of mathematics in its many and varied forms. Distinguished authors will provide an up-to-date summary of the results of current research in their fields and give their own take on what they believe are the most significant debates influencing research, drawing original conclusions.

Cambridge Elements ≡

The Philosophy of Mathematics

Elements in the Series

A full series listing is available at: www.cambridge.org/EPM